T0302103

Smart Nanomaterials

Smart nanomaterials are the basis of diverse emerging applications, and this book covers their technological advances, innovations, and practical applications. It covers advances in the most critical aspects in chemistry and material fabrication of nanomaterials including engineering/prospective applications. The application of smart nanomaterials in the biomedical field, agriculture, food industry, and apparel industry is covered with practical examples. It discusses the future of smart nanomaterials and the pros and cons associated with smart nanomaterials in a detailed manner.

Features:

- Reviews synthesis methods, characterization techniques, and applications of smart nanomaterials.
- Explores the significance of using smart nanomaterials in futuristic life including examples.
- Covers bionsensor sensitivity, selectivity, and stability for long-term operation and nanotechnology technique for possibly detecting hazardous viruses.
- Provides the manufacturing advantages, energy and electronic advantages, medical benefits, environmental effects, and economic issues.
- Explains risk assessment on humans in detail.

This book is aimed at researchers and graduate students in nanomaterials and materials science.

Emerging Materials and Technologies

Series Editor: Boris I. Kharissov

The *Emerging Materials and Technologies* series is devoted to highlighting publications centered on emerging advanced materials and novel technologies. Attention is paid to those newly discovered or applied materials with potential to solve pressing societal problems and improve quality of life, corresponding to environmental protection, medicine, communications, energy, transportation, advanced manufacturing, and related areas.

The series takes into account that, under present strong demands for energy, material, and cost savings, as well as heavy contamination problems and worldwide pandemic conditions, the area of emerging materials and related scalable technologies is a highly interdisciplinary field, with the need for researchers, professionals, and academics across the spectrum of engineering and technological disciplines. The main objective of this book series is to attract more attention to these materials and technologies and invite conversation among the international R&D community.

Application of Numerical Methods in Civil Engineering Problems
M.S.H. Al-Furjan, M. Rabani Bidgoli, Reza Kolahchi, A. Farrokhian, and M.R. Bayati

Advanced Functional Metal-Organic Frameworks: Fundamentals and Applications
Edited by Jay Singh, Nidhi Goel, Ranjana Verma and Ravindra Pratap Singh

Nanoparticles in Diagnosis, Drug Delivery and Nanotherapeutics
Edited by Divya Bajpai Tripathy, Anjali Gupta, Arvind Kumar Jain, Anuradha Mishra and Kuldeep Singh

Functional Nanomaterials for Sensors
Suresh Sagadevan and Won-Chun Oh

Functional Biomaterials: Advances in Design and Biomedical Applications
Anuj Kumar, Durgalakshmi Dhinasekaran, Irina Savina, and Sung Soo Han

Smart Nanomaterials
Imalka Munaweera and M. L. Chamalki Madhusha

For more information about this series, please visit: www.routledge.com/Emerging-Materials-and-Technologies/book-series/CRCEMT

Smart Nanomaterials

Imalka Munaweera and
M. L. Chamalki Madhusha

CRC Press is an imprint of the
Taylor & Francis Group, an **informa** business

First edition published 2024
by CRC Press
6000 Broken Sound Parkway NW, Suite 300, Boca Raton, FL 33487-2742

and by CRC Press
4 Park Square, Milton Park, Abingdon, Oxon, OX14 4RN

CRC Press is an imprint of Taylor & Francis Group, LLC

Library of Congress Cataloging-in-Publication Data
Names: Munaweera, Imalka, author. | Madhusha, M. L. Chamalki, author.
Title: Smart nanomaterials / Imalka Munaweera, M. L. Chamalki Madhusha.
Description: Boca Raton : CRC Press, 2024. | Series: Emerging materials and technologies | Includes bibliographical references and index.
Identifiers: LCCN 2023003726 (print) | LCCN 2023003727 (ebook) |
ISBN 9781032416175 (hardback) | ISBN 9781032432243 (paperback) |
ISBN 9781003366270 (ebook)
Subjects: LCSH: Nanostructured materials. | Smart materials.
Classification: LCC TA418.9.N35 M863 2024 (print) | LCC TA418.9.N35 (ebook) |
DDC 620.1/15—dc23/eng/20230505
LC record available at https://lccn.loc.gov/2023003726
LC ebook record available at https://lccn.loc.gov/2023003727

ISBN: 9781032416175 (hbk)
ISBN: 9781032432243 (pbk)
ISBN: 9781003366270 (ebk)

DOI: 10.1201/9781003366270

Typeset in Times
by codeMantra

Contents

Authors

Dr. Imalka Munaweera is a winner of the 2021 OWSD-Elsevier Foundation Award for Early-Career Women Scientists in the Developing World for her research in Chemistry, Mathematics, and Physics especially for her research contribution to the nanotechnology-related projects. Currently, she is a Senior Lecturer at the Department of Chemistry at the University of Sri Jayewardenepura in Sri Lanka. She has over 10+ years of teaching experience in Nanotechnology, Application of Nanotechnology, Inorganic Chemistry, Polymer Chemistry, and Instrumental Analysis. She has over 15+ years of research experience in the field of Nanotechnology, Inorganic Chemistry, and Materials Science. She obtained her PhD in Chemistry at The University of Texas at Dallas, USA in 2015. She has served as an Assistant Professor in Chemistry, Prairie View A&M University, Prairie View, Texas, USA. She was a Postdoctoral Researcher at the University of Texas Southwestern Medical Center, Dallas, Texas, USA. Her research interests are nanotechnology for drug delivery/pharmaceutical applications, agricultural applications, and water purification applications. Furthermore, she also pursues research toward development of nanomaterials from natural resources for various industrial applications. She has authored many publications in indexed journals and is also an inventor of US-granted patents (licensed and commercialized), international patents, and Sri Lankan-granted patents related to nanoscience and nanotechnology-based research. Further, she is a recipient of many awards related to nanoscience and nanotechnology research. In addition, she is a principal investigator (PI) for a research grant which was awarded by World Academy of Sciences (TWAS) and a core PI for two more grants (NRC PPP-Sri Lanka and TWAS). Apart from being awarded many accolades in both the USA and Sri Lanka, including scholarships, she has also taken part in paper review and many conference presentations and contributed to abstract awards as well. Some of her awards include the National Science & Technology Award in Sri Lanka in 2010 under the category of innovative advanced technologies with commercial potential that was awarded by the President of Sri Lanka and several graduate competition awards awarded by the American Chemical Society.

Ms. M. L. Chamalki Madhusha is a graduate research assistant, and she has obtained her BSc degree in Chemistry from the University of Sri Jayewardenepura, Sri Lanka. Currently, she is working toward a sustainable future through green chemistry and nanotechnology-based findings. Her research interests include materials chemistry, food chemistry, green chemistry, and nanotechnology with high-impact journal publications. She is engaged in generating new research ideas and devising feasible solutions to broadly relevant problems. She is an author of eight indexed publications, two local and international patents, and four international conference abstracts.

Preface

There is a tremendous development in nanotechnology, and rising challenges in almost all aspects such as agriculture, food industry, textile industry, catalytic applications, biomedical applications, and environmental problems have fueled a stringent requirement for developing novel materials. Over the past decades, most of the scientists have been working hard to meet this stringent requirement. Sophisticated materials are now assessable, leading to significant progress in most of the aforementioned industries. Prominent among them are smart nanomaterials, which combine both the excellent properties of nanomaterials with smart functional materials that have caused profound revolution in the understanding of the basic concept of 'smart nanomaterials'.

Impressive progress has been made in this smart nanomaterial fields due to the employment of novel synthesis methods. The use of smart nanomaterials in the wide range of applications enables to do tedious tasks in a simple and smart manner. The unique electronic, magnetic, acoustic, and light properties of nanomaterials, coupled with smart materials capable of responding to external stress, electric and magnetic fields, temperature, moisture, and pH make accurate, real-time, and modulated analysis possible. This book summarizes the main applications of smart nanomaterials in the various fields. The emphasis is to highlight the latest and significant progress made in these fields. We hope to provide insight into some new directions such as smart nanomaterials in renewable technologies and biogenic nanomaterials as well. As such, this book can be used not only as a textbook for undergraduate and graduate students, but also as a reference book for researchers in biotechnology, nanotechnology, biomaterials, medicine, bioengineering, etc.

This approach is rooted in the literature, with copious references, presented at the end of each chapter for ease of access. Our hope is that the content of this book will be sufficient to give a basic understanding of smart nanomaterials and their applications in various aspects such as food industry, biomedical industry, agricultural industry, textile industry, catalysis applications, etc. More importantly, readers will follow the references for details and examples of those aspects in which they are most interested. The abstracts and the key references, at the end of each chapter, are intended to convey the basic messages in a concise way.

Dr. Imalka Munaweera

INTRODUCTION

With the advancement of material science and nanotechnology, many new, high-quality, and cost-effective materials have come into use in a variety of engineering fields. Over the last ten decades, materials have become multifunctional, necessitating the optimization of various characterization and properties. With the previous evolution, the concept has been driving toward composite materials, and recently, the concept of smart materials has been considered as the next evolutionary step. Smart nanomaterials are next-generation materials that outperform traditional structural and functional materials. These materials have inherent intelligence and can adapt to external stimuli such as loads or the environment.

Smart nanomaterials are defined as materials that have the ability to change their physical properties in response to a specific stimulus input. Smart nanomaterials can be employed in a wide variety of fields such as agriculture and food technology, catalysis, biomedical applications, apparels, and fertilizers. In this regard, application of smart nanomaterials plays a significant role in contemporary life. A large number of smart nanomaterials in various aspects with their respective applications have been discussed in detail in this book. In summary, this book offers a broad content on the smart nanomaterials and their applications.

The book offers comprehensive coverage of the most essential topics, including the following:

1. Provides a comprehensive understanding about basic characterization and synthesis methods of smart nanomaterials.
2. The significance of using smart nanomaterials in futuristic life is discussed in detail with suitable examples.
3. Provides a comprehensive understanding about smart nanomaterials in biomedical applications, pharmaceutical applications and analysis, agricultural applications, food preservation, apparel industry, and catalysis and renewable energy applications.
4. Addresses the manufacturing advantages, energy and electronic advantages, medical benefits, environmental effects and economic issues, and risk assessments on humans in detail.
5. Fulfills the timely need of a book that covers the important synthesis methods, characterization techniques, and applications of smart nanomaterials.

Up to date, there are no other books/book chapters which discuss the applications of smart nanomaterials in a wide range with narrated examples. In a nutshell, this book will be a great asset to undergraduates/early-career scientists/beginners of materials science as it provides a comprehensive and complete understanding of different smart nanomaterials and their respective applications in various aspects, in a short time. Intended audience is based on science education while specifically focusing on undergraduates/graduate students/early scientists and beginners of chemistry, materials chemistry, and nanotechnology and nanoscience.

1 What Are Smart Nanomaterials (SNM)?

1.1 DEFINITIONS FOR SNM

Essentially, smart nanomaterials (SNM) are a new generation of functional materials that surpass traditional structural materials. These materials are called SNM because of their self-awareness, self-adaptation, memory ability, and multiple functions (Gottardo et al., 2021). The self-adaptation features of SNM facilitate them to be used in wider range of applications. Due to their ability to selectively change the physical properties in response to changes in environmental factors (stimulus response), SNM are in great demand today in various industries. The external stimuli factors are temperature, stress, magnetic fields, chemicals, electricity, radiation, acidity, and hydrostatic pressure. Changes can be in particle size and shape, stiffness, constraints, and viscosity. All these changes are responsible for providing different required functions of SNM according to the environmental changes.

To overcome the limitations of common materials and nanomaterials, SNM are widely considered. Earlier, these SNM were often defined as a material that can respond to the surrounding in a timely manner (Thangudu, 2020). After that, the SNM have been defined as a material that can be stimulated by external factors and results in a new kind of functional properties. The examples for the stimuli are temperature, light, magnetic field, electric field, stress, pressure, pH, etc. These stimuli control the applications of the SNM. The smartness of these materials lies in their diverse applications such as structural, aerospace, bionics, mechanical engineering, nanotheranostics, food packaging, fertilizers, environmental technology, and so on. In addition, these SNM are able to detect scratches and cracks on any surfaces even in human skin. In this regard, SNM can be served as diagnostic tools and consequently exhibit self-healing capabilities (self-healing effect) (Camboni et al., 2019). Today, SNM are getting a lot of attention because most applications run in an open environment and are subject to a lot of modifications. Therefore, the engineering, bio-medical, food packaging, textile, and agriculture industries are in great demand for various SNM, related molecules and structures (Zelzer and Ulijn, 2010).

1.2 SYNTHESIS METHODS OF SNM

Tunable properties of nanomaterials such as size, high surface area-to-volume ratio, improved stability, sensitivity, specificity, cost-effective commercial-scale fabrication, and so on help nanomaterials to function as optimal substrates in developing SNM. Today, metal nanoparticles (NPs), magnetic nanostructures, conducting polymer-based NPs, and multi-walled carbon nanotubes are some of the most commonly studied nanoparticulate systems in material development. Surface

DOI: 10.1201/9781003366270-1

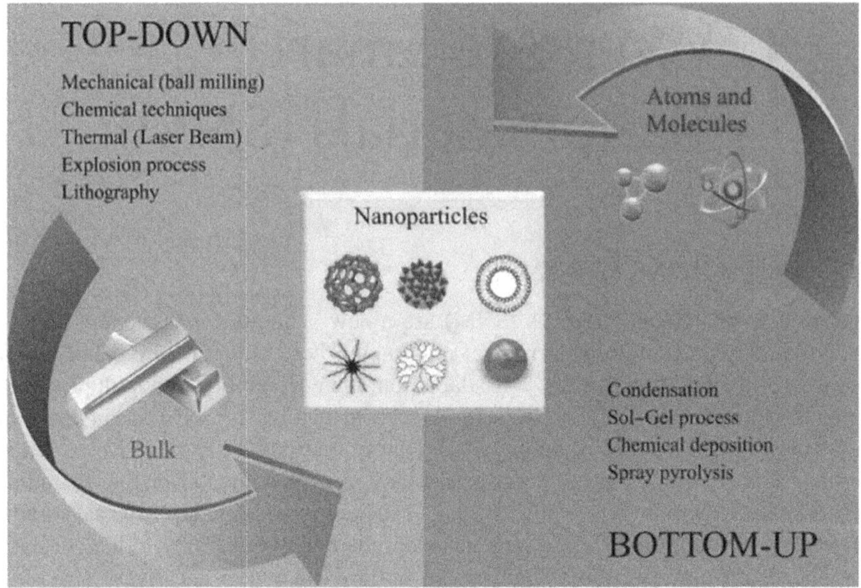

FIGURE 1.1 Top-down and bottom-up approaches exploiting different physical, chemical, and biological methods for the synthesis of nanoparticles. (Barabadi et al., 2019.)

modification of particles with suitable surfactants or polymers protects particles from aggregation, and thus will enhance their stability via functionalization (Ahangaran and Navarchian, 2020). The main two strategies in the synthesis of NPs are as follows: (1) bottom-up approaches and (2) top-down approaches (Figure 1.1). Synthesis of NPs occurs by chemical, physical, or biological strategies, and it depends on the experimental conditions and protocols (Abid et al., 2021).

The top-down synthesis is the decomposition of larger molecules into smaller units and their conversion into suitable NPs via the techniques grinding/milling, chemical vapor deposition (CVD), physical vapor deposition, etc. (Khan et al., 2019). For example, top-down laser fragmentation was reported for the preparation of highly photoactive active Co_3O_4 NPs using laser fragmentation. The irradiations of powerful lasers generate uniform NPs with an average size of 5.8 nm and good oxygen vacancies (Zhou et al., 2016). Mechanochemical milling is also a cost-effective method for producing nanomaterials from bulk materials. Mechanical milling is an effective method for producing blends of different phases, and it is helpful in the production of nanocomposites (Zhuang et al., 2016). Electrospinning is another simple top-down method for the fabrication of nanostructured fibrous materials. It is generally used to fabricate nanofibers from a wide variety of polymeric materials (Munaweera et al., 2014a). Coaxial electrospinning is an effective and simple top-down approach for achieving core–shell ultrathin fibers on a large scale. The lengths of these ultrathin nanomaterials can be extended to several centimeters (Munaweera et al., 2014c). Lithography is another useful tool for developing nanoarchitectures using a focused beam of light or electrons. Lithography can be divided into two main types: masked lithography and maskless lithography (Pimpin and Srituravanich, 2012). In masked nanolithography, nanopatterns are transferred over a large surface area using a specific mask or template. In maskless lithography, arbitrary nanopattern writing is carried out without the involvement of a mask (Baig et al., 2021). Sputtering is a process used to produce nanomaterials via bombarding solid surfaces with high-energy particles such as plasma or gas (Ayyub et al., 2001). Sputtering is considered to be an effective method for synthesis of thin films of nanomaterials (Zhang et al., 2019). The arc discharge method is useful for the generation of various nanostructured materials. It is more known for producing carbon-based materials, such as fullerenes, carbon nano-horns (CNHs), carbon nanotubes, few-layer graphene (FLG), and amorphous spherical carbon NPs. The arc discharge method has great significance in the generation of fullerene nanomaterials. Laser ablation synthesis involves nanoparticle generation using a powerful laser beam that hits the target material (Amendola and Meneghetti, 2009). During the laser ablation process, the source material or precursor vaporizes due to the high energy of the laser irradiation, resulting in nanoparticle formation.

Bottom-up synthesis is named as a building-up approach and is usually performed via (Thiruvengadathan et al., 2013) sol–gel, green synthesis, spinning, biochemical synthesis, etc. A novel synthesis of anatase-type TiO_2 NPs with graphene domains was reported starting from alizarin and titanium isopropoxide precursors for the photocatalytic degradation of methylene blue (Mogilevsky et al., 2014) and synthesis of silica NPs using tetraethyl orthosilicate (TEOS) for various biological applications (Munaweera et al., 2014b).

CVD methods have great significance in the generation of nanomaterials based on carbon. Here, thin films are formed on the various substrate surfaces via the chemical reaction of vapor-phase precursors (Carlsson and Martin, 2010). The hydrothermal process is one of the most well-known and extensively used methods used to produce nanostructured materials and carried out in an aqueous medium at high pressure and temperature around the critical point in a sealed vessel (Li et al., 2015). The solvothermal method is like the hydrothermal method. The only difference is that it is carried out in a non-aqueous medium. The sol–gel method is a wet-chemical technique that is extensively used for the development of various nanomaterials. This method is used for the development of various kinds of high-quality metal-oxide-based nanomaterials. This method is called a sol–gel method as during the synthesis of the metal-oxide NPs, the liquid precursor is transformed to a sol, and the sol is ultimately converted into a network structure that is called a gel. Soft- and hard-template methods are extensively used to produce nanoporous materials. The soft template method is a simple conventional method for the generation of nanostructured materials. The reverse micelle method is also a useful technique for producing nanomaterials with the desired shapes and sizes (Baig et al., 2021).

The nanomaterials that synthesize using above method become to 'smart' by adding functionalities to respond to a stimulus or their environment to produce a dynamic and reversible change in critical properties (Yoshida and Lahann, 2008).

1.3 CHARACTERIZATION OF SNM

A summary of characterization techniques that used in smart nanomaterial characterization is depicted in Figure 1.2.

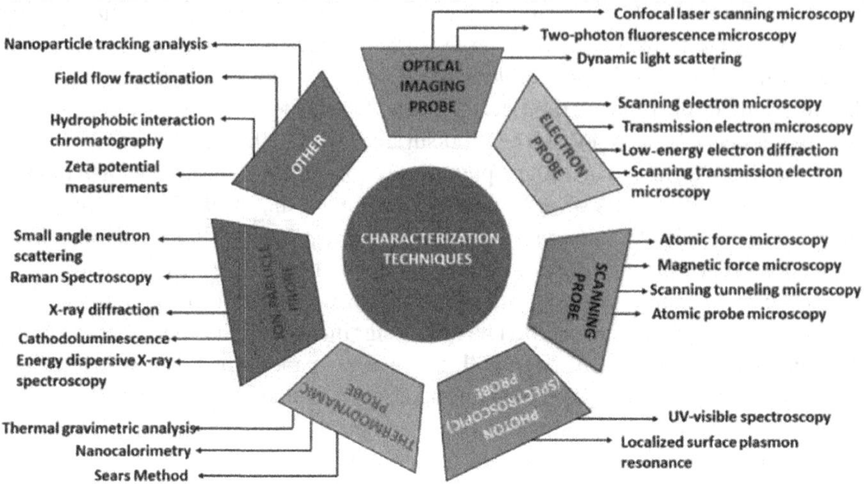

FIGURE 1.2 Different smart nanoparticle characterization techniques. (Sharma and Hussain 2020.)

1.3.1 Morphological Features

The morphological characterization of SNM has a paramount significance because the morphology of nanomaterials profoundly affects majority of the nanomaterials' properties (Bhagyaraj and Oluwafemi, 2018). Morphological features of different nanomaterials vary significantly depending on lattice and crystal structure, matrix composition, synthesis methodology, and thermal stability or photo stability. Furthermore, morphological variation is a facile and effective way of introducing specific functionalities to nanomaterials by affecting their biocompatibility as well (Kumar et al., 2009). As an example, a self-assembling duplex deoxyribonucleic acid (DNA) has been used as building blocks to synthesize three-dimensional DNA structures sized between 10 and 100 nm (Xu et al. 2019a). A specific technique was used to create a different nanoscale "DNA origami" which possesses a different morphology than the used three-dimensional DNA structure (Shin et al., 2020). Therefore, morphological diversity is very significant in determining the functionalities of nanomaterials. Nanomaterials are available in a wide variety of forms, and each one of them is distinctive (Burda et al., 2005). The variety of nanomaterial shapes appears as a result of two factors: the effects of a templating or guiding agent and innate crystallographic growth patterns of the nanomaterials. Spherical micellar emulsion creation is an example of a templating or guiding agent (Glatter and Salentinig, 2020). The other factor that determines the aforementioned morphologies is the innate crystallographic growth patterns of the nanomaterials themselves (Munaweera and Madhusha, 2023). Amorphous particles usually adopt spherical shapes or nanospheres and anisotropic microcrystalline whiskers which correspond to their particular crystal shape (Sayes and Warheit, 2009).

Morphology control of nanomaterials is of utmost importance to effectively explore the features of nanomaterials for usage in a variety of future technologies and applications. Examples of applications based on the optical properties of gold

NPs include optical filters and bio-sensors (Shcikhzadeh et al., 2021). These applications necessitate anisotropy of the particle shape because bigger shapes result in higher plasmon losses (Li et al., 2014). Since well-defined magnetization axis and switching fields are necessary to process or store information, the shape of magnetic NPs cannot be regulated (Gubin, 2009). Furthermore, in order to learn more about the morphological characteristics of NPs, flatness, sphericity, and aspect ratio will be considered in nanomaterials (Sanjay and Pandey, 2017).

Nanowires and nanotubes with various shapes such as belts, helices, zigzags, or nanowires with diameter that varies with length are examples of NPs with high aspect ratios. On the other hand, spherical, oval, cubic, prismatic, helical, or pillar-shaped NPs have smaller aspect ratios (Buzea et al., 2007). These NPs can be found as colloids, suspensions, or solid powders. Additionally, the concept of aspect ratio is the foundation for categorizing nanomaterials based on their dimensionality. For example, surface coatings or thin films are one-dimensional nanomaterials (1D) (Figure 1.3). Two-dimensional (2D) nanomaterials include nanopore filters or 2D

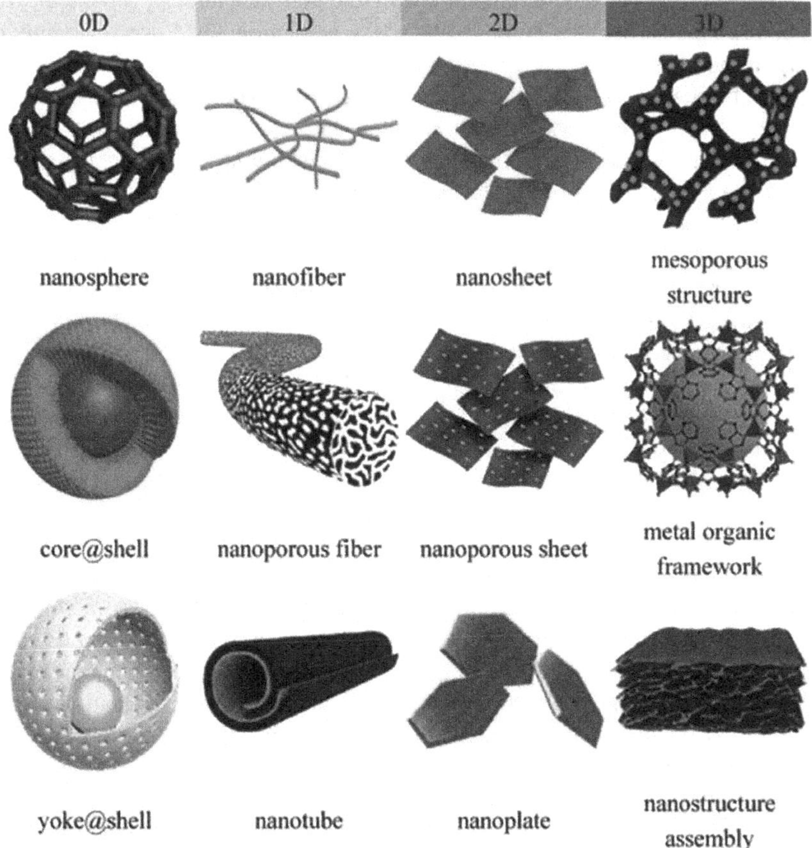

FIGURE 1.3 Representative illustrations of dimensionally different nanomaterials. (Goh et al., 2020.)

nanostructured films with nanostructures securely affixed to a substrate. These materials are composed of thin layers that may have a thickness of at least one atomic layer. Contrary to bulk materials, these nanomaterials have a high aspect ratio (surface-area-to-volume ratio) and therefore have many atoms on their surface (Munaweera and Madhusha, 2023). Three-dimensional (3D) nanomaterials include powders, dispersions of NPs, bundles of nanowires, and nanotubes as well as multi-nanolayers (Paramasivam et al., 2021).

There are different methods to characterize nanomaterials in order to examine the morphology and other parameters, but the most important characterization tools include the use of electron microscopy (Pal et al., 2011). These include scanning electron microscopy (SEM), transmission electron microscopy (TEM), and polarized optical microscopy (POM) (Yang et al., 1997). The SEM is based on the electron examining rule and provides all available information at the nanoscale level about the nanomaterials including the identification of their morphological properties as well as how NPs have scattered within the lattice structure. Activated carbon and hard carbon samples were examined using the SEM technique (Coleman et al., 2006) (Figure 1.4). TEM images highlighted the disordered structures of activated carbon and amorphous nature (Xu et al. 2019b). High-resolution TEM uses a slightly different methodology to other types of TEM, resulting in a high resolution that can structurally characterize samples at an atomic level (Munaweera and Madhusha, 2023). It is, therefore, a very useful technique in the study of all nanoscale structures. Additionally, the SEM technique was used to examine the morphological characteristics of ZnO-doped molecular orbital frameworks (Huang et al., 2019). This approach exposes the morphologies of molecular orbital frameworks and has proven the scattering of ZnO NPs in lattice. Since TEM is reliant on the electron transmission standard, it can reveal several distinctive information about the bulk content when exposed to currents of low to high amplification. The morphologies of gold NPs, for instance, have been studied using TEM (Ong et al., 2017). TEM images of gold NPs have shown certain NPs' properties (Dykman and Khlebtsov, 2011). TEM also provides fundamental information about nanomaterials with layered structures, such as in the Co_3O_4 NPs and activated carbon (Figure 1.4). In Co_3O_4 NPs, the empty quadrupolar shell structure is clearly made visible by TEM (Afolalu et al., 2019).

1.3.2 Structural Defects

For the characterization of structural deficiencies in nanomaterials, numerous characterization approaches have been developed. Direct and indirect methods can be used to characterize these procedures. The microscopic methods, such as SEM or TEM, are included in the category of direct procedures. The investigation of grain boundaries using electron backscatter diffraction (EBSD) in the context of the SEM is precise and successful. When it comes to indirect methods, the main techniques that are frequently utilized include electrical resistometry (ER), X-ray line profile analysis (XLPA), and positron annihilation spectroscopy (PAS) (Gubicza, 2017). Characterization is straightforward and simple with techniques like PAS, ER, and XLPA since they are nondestructive techniques. Thin surface layers can be analyzed by utilizing the PAS and XLPA. In this case, the surface of the relevant specimen

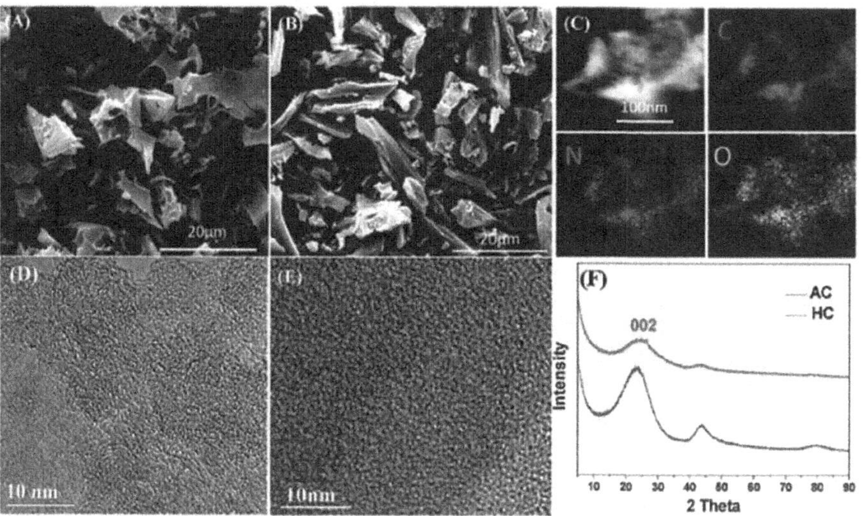

FIGURE 1.4 (a) SEM image of activated carbon "AC". (b) SEM image of hard carbon "HC". (c) TEM HAADF image and EDXS analytical maps of C, N, and O for AC. (d, e) HRTEM images highlighting the disordered structures of AC and HC, respectively. (f) X-ray diffraction patterns of AC and HC, highlighting their amorphous structure. (Xu et al., 2019b.)

must be chopped or machined and removed using electropolishing or etching. In addition, mechanical polishing of nanomaterials is not advised as a suitable surface preparation for PAS or XLPA since it may seriously alter the nanomaterial's surface layer. As a result, the results are connected to the surface fault structure that has been mechanically polished and do not characterize the bulk material. Therefore, this outermost deformed layer must be removed prior to PAS or XLPA analyses if earlier investigations such as hardness assessment required a mechanically polished surface (Munaweera and Madhusha, 2023). EBSD stands out when looking at nondestructive procedures. The most accurate and sensitive method for determining the surface quality of nanomaterials is EBSD. This is brought about by the extremely thin surface layer analyzed in this technique and the method's susceptibility to contamination and surface roughness (Gubicza, 2017). In general, EBSD technology requires a pure, smooth, and distortion-free surface. As a result, there are a number of procedures frequently involved in surface preparation of the particular nanomaterial which are briefly covered here. Mechanical polishing is the first step toward obtaining a smooth surface. Then, electropolishing or ion polishing is required to remove the mechanical polishing-induced distortion in the topmost layer. The most challenging sample preparation is required for TEM, which includes mechanical thinning of the sample to a thickness of 80mm and then jet polishing or ion milling the sample until it is perforated (Mukhopadhyay, 2003). The sample is frequently chilled to liquid nitrogen temperature (77 K) during the thinning by an ion beam in order to prevent the recovery and recrystallization of nanostructures due to the temperature rise produced by the ion beam. Compared to TEM, ER method examines a bigger volume.

Additionally, the volume probed by XLPA or PAS is considerably bigger than that by TEM (Čížek et al., 2019). The larger the analyzed volume of nanomaterial, the better the statistics of the parameters acquired for the defect structures.

1.3.3 Optical Studies

One of the major characteristic properties that govern the photocatalytic application of nanomaterials is their optical property. When considering the optical properties of nanomaterials, the Beer Lambert law plays a significant role, and it states the quantity of light absorbed by a substance dissolved in a fully transmitting solvent is directly proportional to the concentration of the substance and the path length of the light through the solution. The techniques that are used in optical characterization are important to reveal the details of light absorption, reflectivity of nanomaterials, luminescence, and phosphorescence properties. Semiconductor nanomaterials and metallic NPs, for example, are extensively used for photocatalytic applications (Khan et al., 2019).

The instruments that are being used in identification of optical properties are UV–Vis spectrophotometer, photoluminescence (PL) and spectroscopic ellipsometry. The Diffuse Reflectance Spectrometer (DRS) is mainly used for the evaluation of bandgap values and solid UV spectra of solid nanomaterials (Sangiorgi et al., 2017). DRS technique is specific since it is used for only solid materials to measure the band gaps (Munaweera and Madhusha, 2023).

Bandgap evaluation is of paramount importance in evaluation of photo activity and conductivity of nanomaterials. Graphitic carbon nitride (g-C_3N_4) has become a new research hotspot and drawn broad interdisciplinary attention as a metal-free and visible-light-responsive photocatalyst in the arena of solar energy conversion and environmental remediation. Graphitic carbon nitride (g-C_3N_4) possesses a bandgap of ~2.7 eV (~460 nm). The photographic limits of this nanomaterial are explicitly linked to the UV–Vis spectroscopy (2.74–2.77 eV) bandgap estimate. In addition, this framework also provides absorption movement in case of doping event, compound action or hetero-structure of nanomaterials. In order to examine the range of optical properties of $LaFeO_3$, montmorillonite, and $LaFeO_3$/montmorillonite nanocomposites, their UV electromagnet absorption range was investigated (Afolalu et al., 2019). $LaFeO_3$/montmorillonite and $LaFeO_3$ demonstrated a comparatively wide range of absorption between 400 and 620 nm which could be attributed to the electronic transition from the valence band to the conduction band (Munaweera and Madhusha, 2023). The $LaFeO_3$ and $LaFeO_3$/MMT particles could absorb considerable amounts of visible light, which implied their potential applications as visible light-driven photocatalysts (Peng et al., 2016). PL is also considered a fundamental technology for investigating the optical properties of photoactive NPs despite their UV effects (Li and Zhang, 2009). The retention or breakdown of the material's radiation point and its impact on the overall photo excitons of the excitation time are further explained by this method. The PL range may be varied depending on the study. A typical PL range is developed with defect-free and balanced ZnO NPs, and from this range, it is clear that the defect-free ZnO NPs have higher PL control when the ZnO NPs changed to CdS (Eixenberger et al., 2019). The CdS/Au/ZnO composite shows the lowest performance. In the recent past, this damping drop from CdS/Au/ZnO to unaltered ZnO can be attributed to a decrease in the charge recombination rate and a longer photo exciton lifetime (Gurugubelli et al., 2022). Additionally, the doping rate of the material in the NPs, layer thickness, and oxygen opening are all determined using this framework. Wan et al. also used spectroscopic ellipsometry methods to select estimates of refractor reduction and tail coefficients for bare gold NPs (Beaudette, 2021). In order to identify optical constants with various morphologies and plasmonic properties, they organized the motion of gold NPs. Make a distinction between the properties and optical continuous estimates of solid gold NPs, exhibiting outstanding evidence of using these materials in applications for compound identification, given their sensitivity as demonstrated by ellipsometric values (Beaudette, 2021).

1.3.4 Structural Analysis

Structural analysis of nanomaterials is significant since it is a prerequisite for better understanding of the characteristics of nanomaterials to have a detailed knowledge of the structure from the atomic/molecular (local) level to the crystal lattice structure and to the microstructure. The main objectives of structural analysis of nanomaterials are to investigate the relationship between structure and property as well as to reveal novel properties.

A key method for determining the crystallite structure and average crystallite size of nanomaterials is powder X-ray diffraction (PXRD). The morphological and structural details of the researched nanomaterials can be studied using X-ray scattering and Bragg diffraction (Giannini et al., 2016). From the PXRD analysis, a variety of structural details can be obtained including the atomic structure of the crystal, the positions and symmetry of the atoms in the unit cells, the size and shape of the nanocrystalline domain, the identification of the crystalline phases, a quantitative estimation of their weight fractions, and the positions and symmetry of the nanoscale assembly's NPs and nanocrystals as well as the assembly's length (Hens and De Roo, 2020). Additionally, TEM provides a clear, consistent, and straightforward technique to visualize the atomic lattice in a crystalline nanomaterial (Chen et al., 2013). The method's strength and significance are further increased by the ability to reveal minute crystal structure features. On the other hand, a major drawback of the traditional TEM is that the imaging resolution that is theoretically predicted cannot be achieved (Wall and Hainfeld, 1986). The reasons behind this drawback are the inability to control the image-forming electron beam in the instrument and stability of the high-voltage source for the filament or level of vacuum. As a result of the aforementioned reasons, true atomic resolution images have been incomprehensible. With the development of nanoscience, new advances in TEM technology have increased the resolution limit to the sub-angstrom scale (Ercius et al., 2015). By using aberration-corrected scanning transmission electron microscopy (STEM), which relies on raster scanned electron beam over the sample, and a high-angle annular dark field (HAADF) detector, which provides enhanced Z contrast, electron microscopy images with an atomic resolution (nominally 0.8 Å) have been achieved (Sang et al., 2017) . The findings have given us a unique perspective on the intricate structure of NPs and distinctive proof that sophisticated nanomaterials exist (Munaweera and Madhusha, 2023).

1.3.5 Elemental Studies

There are various kinds of medicinal, commercial, and bio-based applications of nanomaterials in the recent era. Regardless of the infinitesimally small size of nanomaterials, they can pose significant risks to human health. In this regard, a detailed and complete analysis of the elemental composition of respective nanomaterials is vital to ensure the safe and sustainable establishment of nanomaterials in human applications. The most trusted and well-known method to analyze the elemental composition of nanomaterials is mass spectrometry (MS) (Chait, 2011).

Elemental mapping utilizes the compositional accuracy built into methods like EDS microanalysis and combines it with high-resolution imaging to present complex data in an approachable, aesthetically appealing format. The foundation of elemental mapping is the collection of incredibly precise elemental composition data over a sample's surface. Usually, a SEM or TEM is used for this, along with EDS analysis. Along with the EDS data, a high-resolution image of the area of interest is gathered, and the two are then correlated. A full elemental spectrum is also gathered for every pixel in the digital image (Munaweera and Madhusha, 2023).

1.3.6 Size Estimation

The size estimation of NPs can be characterized by different techniques which include TEM, SEM, atomic force microscopy (AFM), X-ray diffraction (XRD), and dynamic light scattering (DLS). Usually, TEM, AFM, XRD, and SEM techniques give better estimation of the size of nanomaterials than DLS measurements because the latter technique estimates the size of nanomaterials at incredibly low dimensions. The DLS technique has been used to study the size variation of silica NPs with the absorption rate of serum protein. The size of the silica NPs has expanded with the protein layer. In this regard, the hydrophilicity of respective NPs and agglomeration of NPs affected the DLS measurements (Mourdikoudis et al., 2018). Therefore, the differential centrifugal sedimentation (DCS) technique is considered in such situations (Langevin et al., 2018). For the investigation of DNA, proteins, and other organic compounds, a rather strong and distinctive methodology is available and it helps to track NPs in addition to the DCS method (Yang et al. 2014). Furthermore, the size distribution profile of the NP in a fluid medium with diameters ranging from 10 to 1000 nm can be located by correlating the rate of Brownian motion to the particle size. This is done by using the Nanoparticle tracking analysis (NTA) strategy to analyze NPs in the liquid medium (Afolalu et al. 2019). Compared to DLS measurements, the NTA technique interprets more accurate and precise results with far better peak resolution when used in sizing monodispersed and poly-dispersed samples (Oktaviani, 2021).

The large surface area of nanomaterials provides exceptional and extraordinary performance for a wide variety of applications, while the Brunauer–Emmett–Teller (BET) method is the most widely used technique to measure the surface area (Munaweera and Madhusha, 2023). This technique is based upon the BET hypothesis and the adsorption and desorption rules (Tian and Wu, 2018).

1.3.7 Physicochemical Characteristics

Physicochemical properties of nanomaterials include size and shape of the nanomaterials, solubility in different solvent systems, surface area measurements, chemical composition, shape, agglomeration state, crystallinity, surface energy, surface charge, surface morphology, and surface coating (Gatoo et al., 2014). Moreover, the electronic and optical characteristics of nanomaterials play a major role in most of the applications, and hence, the characterization of the physicochemical properties of nanomaterials is essential. As an example, metallic nanomaterials exhibit collective oscillation bands of electrons excited by the incident photons at the resonant frequency, and we can clearly see these bands by a UV spectrum. However, these oscillation bands are apparently absent in mass metal scope, and such observations can be exploited by localized surface plasmon resonance (LSPR) (Singh and Strouse, 2010). The excitation of LSPR bands enables the tunable and enhanced electromagnetic fields, light absorption, and scattering based on the physical and elemental parameters of nanomaterials which in turn upgrades the performances. The LSPR spectrum interprets the shape, dimensions, and interparticle distances, which differentiates the nanomaterials from dielectric properties including solvents, adsorbents, and substrates (Gatoo et al., 2014). As an example, the mean free path for silver metal is roughly 50 nm. The LSPR spectrum of silver shows three distinct bands corresponding to the in-plane dipole, quadrupole, and out-plane quadrupole plasmon resonance (Wu et al., 2015). The SPR shifted to shorter wavelengths, and as a result of that, the edge length of the silver NPs has decreased (Munaweera and Madhusha, 2023).

1.3.8 Magnetic Properties

Magnetic properties of nanomaterials have paramount importance in most of the applications in modern nanotechnology. Therefore, the estimation of magnetic properties of nanomaterials is vital (Kolhatkar et al., 2013). The magnetic characteristics of NPs generally rely on size, and nanomaterials can be used for the synthesis and design of materials with desired magnetic properties based on the size and structure (Figure 1.5).

Nanomaterials are made up of large magnetic particles that are typically divided into a number of magnetic domains with distinct magnetization directions. Small magnetic particles in certain nanomaterials with dimensions less than a threshold size, on the other hand, are made up of a single domain. Magnetization in single-domain particles may be stable; however, magnetization in very small particles is unstable above the superparamagnetic blocking temperature (Dunlop, 1973). This is due to the fact that the energy required to reverse the magnetization is proportional to the volume, and the thermal energy may then be sufficient to result in superparamagnetic relaxation, such as changes in the magnetization direction of extremely small particles. Superparamagnetic relaxation renders the particles unsuitable for magnetic data storage applications, but it may be required at maximum performance in many other applications, such as magnetic beads used in biotechnology (Houshiar et al., 2014). Superparamagnetic relaxation can be explored using a variety of timescale-dependent techniques, including dc and ac magnetization measurements, Mossbauer spectroscopy, and neutron scattering (Mørup

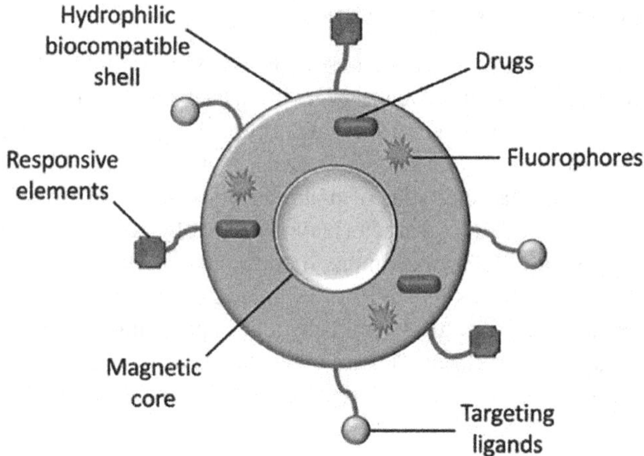

FIGURE 1.5 A classic illustration of a magnetic nanoparticle's structure. (Alromi et al., 2021.)

et al., 2007). As a result, these techniques can be utilized to investigate superparamagnetic relaxation over a wide variety of relaxation times (Munaweera and Madhusha, 2023). Magnetic fields can be used to suppress superparamagnetic relaxation. Strong interparticle interactions can potentially cause superparamagnetic relaxation to be suppressed. When considering magnetic dynamics, it is dominated by excitation of the uniform spin-wave mode, which exhibits a linear relationship between temperature and magnetization (Fazlali et al., 2016). More importantly, the surface effects also dominate with the variation of magnetic properties of nanomaterials. With the increment of temperature, the surface magnetization decreases rapidly. As a result, magnetization in the interior of a particle will be lower than its surface magnetization. The magnetic anisotropy of nanomaterials will be prominently governed by the low symmetry around surface atoms. Moreover, the magnetic properties in the nanomaterial's surface and the defects in the interior can be profoundly influenced by a reduced number of magnetic neighbor atoms, and ultimately this can lead to non-collinear spin structures in ferrimagnetic particles (Issa et al., 2013).

REFERENCES

Abid, Namra, Aqib Muhammad Khan, Sara Shujait, Kainat Chaudhary, Muhammad Ikram, Muhammad Imran, Junaid Haider, Maaz Khan, Qasim Khan, and Muhammad Maqbool. 2021. "Synthesis of nanomaterials using various top-down and bottom-up approaches, influencing factors, advantages, and disadvantages: A review." *Advances in Colloid and Interface Science* no. 300:102597.

Afolalu, Adeniran Sunday, Samuel B Soetan, Samson O Ongbali, Abiodun A Abioye, and AS Oni. 2019. Morphological characterization and physio-chemical properties of nanoparticle-review. Paper read at IOP Conference Series: Materials Science and Engineering.

Ahangaran, Fatemeh, and Amir H Navarchian. 2020. "Recent advances in chemical surface modification of metal oxide nanoparticles with silane coupling agents: A review." *Advances in Colloid and Interface Science* no. 286:102298.

Alromi, Dalal A, Seyed Yazdan Madani, and Alexander Seifalian. 2021. "Emerging application of magnetic nanoparticles for diagnosis and treatment of cancer." *Polymers* no. 13 (23):4146.

Amendola, Vincenzo, and Moreno Meneghetti. 2009. "Laser ablation synthesis in solution and size manipulation of noble metal nanoparticles." *Physical Chemistry Chemical Physics* no. 11 (20):3805–3821.

Ayyub, Pushan, Ramesh Chandra, P Taneja, AK Sharma, and R Pinto. 2001. "Synthesis of nanocrystalline material by sputtering and laser ablation at low temperatures." *Applied Physics A* no. 73 (1):67–73.

Baig, Nadeem, Irshad Kammakakam, and Wail Falath. 2021. "Nanomaterials: A review of synthesis methods, properties, recent progress, and challenges." *Materials Advances* no. 2 (6):1821–1871.

Barabadi, Hamed, Masoud Najafi, Hadi Samadian, Asaad Azarnezhad, Hossein Vahidi, Mohammad Ali Mahjoub, Mahbobeh Koohiyan, and Amirhossein Ahmadi. 2019. "A systematic review of the genotoxicity and antigenotoxicity of biologically synthesized metallic nanomaterials: Are green nanoparticles safe enough for clinical marketing?" *Medicina* no. 55 (8):439.

Beaudette, Chad A. 2021. *Nanostructures, nanoparticles, and 2D materials from nonthermal plasmas*. University of Minnesota, Minneapolis, MO.

Bhagyaraj, Sneha Mohan, and Oluwatobi Samuel Oluwafemi. 2018. "Nanotechnology: The science of the invisible." In Sneha Mohan Bhagyaraj, Oluwatobi Samuel Oluwafemi, Nandakumar Kalarikkal, Sabu Thomas (Eds.), *Synthesis of inorganic nanomaterials*, 1–18. Elsevier, Sawston.

Burda, Clemens, Xiaobo Chen, Radha Narayanan, and Mostafa A El-Sayed. 2005. "Chemistry and properties of nanocrystals of different shapes." *Chemical Reviews* no. 105 (4):1025–1102.

Buzea, Cristina, Ivan I Pacheco, and Kevin Robbie. 2007. "Nanomaterials and nanoparticles: Sources and toxicity." *Biointerphases* no. 2 (4):MR17-MR71.

Camboni, Marco, James Hanlon, Rodrigo Pérez-García, and Pete Floyd. 2019. "A state of play study of the market for so-called "next generation" nanomaterials."

Carlsson, Jan-Otto, and Peter M Martin. 2010. "Chemical vapor deposition." In Peter M. Martin (Ed.), *Handbook of deposition technologies for films and coatings*, 314–363. Elsevier, Norwich, NY.

Chait, Brian T. 2011. "Mass spectrometry in the postgenomic era." *Annual Review of Biochemistry* no. 80 (1):239–246.

Chen, Chien-Chun, Chun Zhu, Edward R White, Chin-Yi Chiu, MC Scott, BC Regan, Laurence D Marks, Yu Huang, and Jianwei Miao. 2013. "Three-dimensional imaging of dislocations in a nanoparticle at atomic resolution." *Nature* no. 496 (7443):74–77.

Čížek, Jan, Miloš Janeček, T Vlasák, B Smola, O Melikhova, RK Islamgaliev, and Sergey Vladimir Dobatkin. 2019. "The development of vacancies during severe plastic deformation." *Materials Transactions* no. 60 (8):1533–1542.

Coleman, Jonathan N, Umar Khan, and Yurii K Gun'ko. 2006. "Mechanical reinforcement of polymers using carbon nanotubes." *Advanced Materials* no. 18 (6):689–706.

Dunlop, David J. 1973. "Superparamagnetic and single-domain threshold sizes in magnetite." *Journal of Geophysical Research* no. 78 (11):1780–1793.

Dykman, Lev A, and Nikolai G Khlebtsov. 2011. "Gold nanoparticles in biology and medicine: Recent advances and prospects." *Acta Naturae (англоязычная версия)* no. 3:34–55.

Eixenberger, Josh E, Catherine B Anders, Katelyn Wada, Kongara M Reddy, Raquel J Brown, Jonathan Moreno-Ramirez, Ariel E Weltner, Chinnathambi Karthik, Dmitri A Tenne, and Daniel Fologea. 2019. "Defect engineering of ZnO nanoparticles for bioimaging applications." *ACS Applied Materials & Interfaces* no. 11 (28):24933–24944.

Ercius, Peter, Osama Alaidi, Matthew J Rames, and Gang Ren. 2015. "Electron tomography: A three-dimensional analytic tool for hard and soft materials research." *Advanced Materials* no. 27 (38):5638–5663.

Fazlali, Masoumeh, Mykola Dvornik, Ezio Iacocca, Philipp Dürrenfeld, Mohammad Haidar, Johan Åkerman, and Randy K Dumas. 2016. "Homodyne-detected ferromagnetic resonance of in-plane magnetized nanocontacts: Composite spin-wave resonances and their excitation mechanism." *Physical Review B* no. 93 (13):134427.

Gatoo, Manzoor Ahmad, Sufia Naseem, Mir Yasir Arfat, Ayaz Mahmood Dar, Khusro Qasim, and Swaleha Zubair. 2014. "Physicochemical properties of nanomaterials: Implication in associated toxic manifestations." *BioMed Research International* no. 2014:498420.

Giannini, Cinzia, Massimo Ladisa, Davide Altamura, Dritan Siliqi, Teresa Sibillano, and Liberato De Caro. 2016. "X-ray diffraction: A powerful technique for the multiple-length-scale structural analysis of nanomaterials." *Crystals* no. 6 (8):87.

Glatter, Otto, and Stefan Salentinig. 2020. "Inverting structures: From micelles via emulsions to internally self-assembled water and oil continuous nanocarriers." *Current Opinion in Colloid & Interface Science* no. 49:82–93.

Goh, Pei Sean, Kar Chun Wong, and Ahmad Fauzi Ismail. 2020. "Nanocomposite membranes for liquid and gas separations from the perspective of nanostructure dimensions." *Membranes* no. 10 (10):297.

Gottardo, Stefania, Agnieszka Mech, Jana Drbohlavová, Aleksandra Małyska, Søren Bøwadt, Juan Riego Sintes, and Hubert Rauscher. 2021. "Towards safe and sustainable innovation in nanotechnology: State-of-play for smart nanomaterials." *NanoImpact* no. 21:100297.

Gubicza, Jeno. 2017. *Defect structure and properties of nanomaterials*. Woodhead Publishing, Sawston.

Gubin, Sergey P. 2009. *Magnetic nanoparticles*. John Wiley & Sons, Weinheim.

Gurugubelli, Thirumala Rao, RVSSN Ravikumar, and Ravindranadh Koutavarapu. 2022. "Enhanced photocatalytic activity of ZnO–CdS composite nanostructures towards the degradation of rhodamine B under solar light." *Catalysts* no. 12 (1):84.

Hens, Zeger, and Jonathan De Roo. 2020. "Atomically precise nanocrystals." *Journal of the American Chemical Society* no. 142 (37):15627–15637.

Houshiar, Mahboubeh, Fatemeh Zebhi, Zahra Jafari Razi, Ali Alidoust, and Zohreh Askari. 2014. "Synthesis of cobalt ferrite ($CoFe_2O_4$) nanoparticles using combustion, coprecipitation, and precipitation methods: A comparison study of size, structural, and magnetic properties." *Journal of Magnetism and Magnetic Materials* no. 371:43–48.

Huang, Lei, Zhihui Yang, Yujun Shen, Peng Wang, Baocheng Song, Yingjie He, Weichun Yang, Haiying Wang, Zhenxing Wang, and Yongsheng Chen. 2019. "Organic frameworks induce synthesis and growth mechanism of well-ordered dumbbell-shaped ZnO particles." *Materials Chemistry and Physics* no. 232:129–136.

Issa, Bashar, Ihab M Obaidat, Borhan A Albiss, and Yousef Haik. 2013. "Magnetic nanoparticles: Surface effects and properties related to biomedicine applications." *International Journal of Molecular Sciences* no. 14 (11):21266–21305.

Khan, Ibrahim, Khalid Saeed, and Idrees Khan. 2019. "Nanoparticles: Properties, applications and toxicities." *Arabian Journal of Chemistry* no. 12 (7):908–931.

Kolhatkar, Arati G, Andrew C Jamison, Dmitri Litvinov, Richard C Willson, and T Randall Lee. 2013. "Tuning the magnetic properties of nanoparticles." *International Journal of Molecular Sciences* no. 14 (8):15977–16009.

Kumar, Annamalai Pratheep, Dilip Depan, Namrata Singh Tomer, and Raj Pal Singh. 2009. "Nanoscale particles for polymer degradation and stabilization—trends and future perspectives." *Progress in Polymer Science* no. 34 (6):479–515.

Langevin, Dominique, Omar Lozano, Anna Salvati, Vikram Kestens, Marco Monopoli, Eric Raspaud, Sandrine Mariot, Anniina Salonen, Steffi Thomas, and Marc D Driessen. 2018.

"Inter-laboratory comparison of nanoparticle size measurements using dynamic light scattering and differential centrifugal sedimentation." *NanoImpact* no. 10:97–107.

Li, Jianlin, Qingliu Wu, and Ji Wu. 2015. *Handbook of nanoparticles*, 1–28. Springer International Publishing, Cham.

Li, Jinghong, and Jin Z Zhang. 2009. "Optical properties and applications of hybrid semiconductor nanomaterials." *Coordination Chemistry Reviews* no. 253 (23–24):3015–3041.

Li, Na, Pengxiang Zhao, and Didier Astruc. 2014. "Anisotropic gold nanoparticles: Synthesis, properties, applications, and toxicity." *Angewandte Chemie International Edition* no. 53 (7):1756–1789.

Mogilevsky, Gregory, Olga Hartman, Erik D Emmons, Alex Balboa, Jared B DeCoste, Bryan J Schindler, Ivan Iordanov, and Christopher J Karwacki. 2014. "Bottom-up synthesis of anatase nanoparticles with graphene domains." *ACS Applied Materials & Interfaces* no. 6 (13):10638–10648.

Mørup, Steen, Daniel E Madsen, Cathrine Frandsen, Christian RH Bahl, and Mikkel F Hansen. 2007. "Experimental and theoretical studies of nanoparticles of antiferromagnetic materials." *Journal of Physics: Condensed Matter* no. 19 (21):213202.

Mourdikoudis, Stefanos, Roger M Pallares, and Nguyen TK Thanh. 2018. "Characterization techniques for nanoparticles: Comparison and complementarity upon studying nanoparticle properties." *Nanoscale* no. 10 (27):12871–12934.

Mukhopadhyay, Sharmila M. 2003. "Sample preparation for microscopic and spectroscopic characterization of solid surfaces and films." In Somenath Mitra (Ed.), *Sample preparation techniques in analytical chemistry*, 377–412. John Wiley & Sons, Ltd., Hoboken, NJ.

Munaweera, Imalka, Ali Aliev, and Kenneth J Balkus Jr. 2014a. "Electrospun cellulose acetate-garnet nanocomposite magnetic fibers for bioseparations." *ACS Applied Materials & Interfaces* no. 6 (1):244–251.

Munaweera, Imalka, Bhuvaneswari Koneru, Yi Shi, Anthony J Di Pasqua, and Kenneth J Balkus Jr. 2014b. "Chemoradiotherapeutic wrinkled mesoporous silica nanoparticles for use in cancer therapy." *APL Materials* no. 2 (11):113315.

Munaweera, Imalka, Daniel Levesque-Bishop, Yi Shi, Anthony J Di Pasqua, and Kenneth J Balkus Jr. 2014c. "Radiotherapeutic bandage based on electrospun polyacrylonitrile containing holmium-166 iron garnet nanoparticles for the treatment of skin cancer." *ACS Applied Materials & Interfaces* no. 6 (24):22250–22256.

Munaweera, Imalka and Madhusha, M.L. Chamalki. 2023. *Characterization techniques for nanomaterials*. CRC Press, Boca Raton, FL.

Oktaviani, Oktaviani. 2021. "Nanoparticles: Properties, applications and toxicities." *Jurnal Latihan* no. 1 (2):11–20.

Ong, Quy, Zhi Luo, and Francesco Stellacci. 2017. "Characterization of ligand shell for mixed-ligand coated gold nanoparticles." *Accounts of Chemical Research* no. 50 (8):1911–1919.

Pal, Sovan Lal, Utpal Jana, Prabal Kumar Manna, Guru Prasad Mohanta, and R Manavalan. 2011. "Nanoparticle: An overview of preparation and characterization." *Journal of Applied Pharmaceutical Science* no. 1:228–234.

Paramasivam, Gokul, Vishnu Vardhan Palem, Thanigaivel Sundaram, Vickram Sundaram, Somasundaram Chandra Kishore, and Stefano Bellucci. 2021. "Nanomaterials: Synthesis and applications in theranostics." *Nanomaterials* no. 11 (12):3228.

Peng, Kang, Liangjie Fu, Huaming Yang, and Jing Ouyang. 2016. "Perovskite $LaFeO_3$/montmorillonite nanocomposites: Synthesis, interface characteristics and enhanced photocatalytic activity." *Scientific Reports* no. 6 (1):1–10.

Pimpin, Alongkorn, and Werayut Srituravanich. 2012. "Review on micro-and nanolithography techniques and their applications." *Engineering Journal* no. 16 (1):37–56.

Sang, Xiahan, Andrew R Lupini, Jilai Ding, Sergei V Kalinin, Stephen Jesse, and Raymond R Unocic. 2017. "Precision controlled atomic resolution scanning transmission electron microscopy using spiral scan pathways." *Scientific Reports* no. 7 (1):1–12.

Sangiorgi, Nicola, Lucrezia Aversa, Roberta Tatti, Roberto Verucchi, and Alessandra Sanson. 2017. "Spectrophotometric method for optical band gap and electronic transitions determination of semiconductor materials." *Optical Materials* no. 64:18–25.

Sanjay, Sharda Sundaram, and Avinash C Pandey. 2017. "A brief manifestation of nanotechnology." In Ashutosh Kumar Shukla (Ed.), *EMR/ESR/EPR spectroscopy for characterization of nanomaterials*, 47–63. Springer, Berlin/Heidelberg.

Sayes, Christie M, and David B Warheit. 2009. "Characterization of nanomaterials for toxicity assessment." *Wiley Interdisciplinary Reviews: Nanomedicine and Nanobiotechnology* no. 1 (6):660–670.

Sharma, Deepali, and Chaudhery Mustansar Hussain. 2020. "Smart nanomaterials in pharmaceutical analysis." *Arabian Journal of Chemistry* no. 13 (1):3319–3343.

Sheikhzadeh, Elham, Valerio Beni, and Mohammed Zourob. 2021. "Nanomaterial application in bio/sensors for the detection of infectious diseases." *Talanta* no. 230:122026.

Shin, Seung Won, Ji Soo Yuk, Sang Hun Chun, Yong Taik Lim, and Soong Ho Um. 2020. "Hybrid material of structural DNA with inorganic compound: Synthesis, applications, and perspective." *Nano Convergence* no. 7 (1):1–12.

Singh, Mani Prabha, and Geoffrey F Strouse. 2010. "Involvement of the LSPR spectral overlap for energy transfer between a dye and Au nanoparticle." *Journal of the American Chemical Society* no. 132 (27):9383–9391.

Thangudu, Suresh. 2020. "Next generation nanomaterials: Smart nanomaterials, significance, and biomedical applications." In Firdos Alam Khan (Ed.), *Applications of nanomaterials in human health*, 287–312. Springer, Berlin/Heidelberg.

Thiruvengadathan, Rajagopalan, Venumadhav Korampally, Arkasubhra Ghosh, Nripen Chanda, Keshab Gangopadhyay, and Shubhra Gangopadhyay. 2013. "Nanomaterial processing using self-assembly-bottom-up chemical and biological approaches." *Reports on Progress in Physics* no. 76 (6):066501.

Tian, Yun, and Jianzhong Wu. 2018. "A comprehensive analysis of the BET area for nanoporous materials." *AIChE Journal* no. 64 (1):286–293.

Wall, JS, and JF Hainfeld. 1986. "Mass mapping with the scanning transmission electron microscope." *Annual Review of Biophysics and Biophysical Chemistry* no. 15 (1):355–376.

Wu, Chunfang, Xue Zhou, and Jie Wei. 2015. "Localized surface plasmon resonance of silver nanotriangles synthesized by a versatile solution reaction." *Nanoscale Research Letters* no. 10 (1):1–6.

Xu, Yang, Shuoxing Jiang, Chad R Simmons, Raghu Pradeep Narayanan, Fei Zhang, Ann-Marie Aziz, Hao Yan, and Nicholas Stephanopoulos. 2019a. "Tunable nanoscale cages from self-assembling DNA and protein building blocks." *ACS Nano* no. 13 (3):3545–3554.

Xu, Ziqiang, Mengqiang Wu, Zhi Chen, Cheng Chen, Jian Yang, Tingting Feng, Eunsu Paek, and David Mitlin. 2019b. "Direct structure–performance comparison of all-carbon potassium and sodium ion capacitors." *Advanced Science* no. 6 (12):1802272.

Yang, Dennis T, Xiaomeng Lu, Yamin Fan, and Regina M Murphy. 2014. "Evaluation of nanoparticle tracking for characterization of fibrillar protein aggregates." *AIChE Journal* no. 60 (4):1236–1244.

Yang, Hong, Neil Coombs, and Geoffrey A Ozin. 1997. "Morphogenesis of shapes and surface patterns in mesoporous silica." *Nature* no. 386 (6626):692–695.

Yoshida, Mutsumi, and Joerg Lahann. 2008. "Smart nanomaterials." *ACS Nano* no. 2 (6):1101–1107.

Zelzer, Mischa, and Rein V Ulijn. 2010. "Next-generation peptide nanomaterials: Molecular networks, interfaces and supramolecular functionality." *Chemical Society Reviews* no. 39 (9):3351–3357.

Zhang, Da, Kai Ye, Yaochun Yao, Feng Liang, Tao Qu, Wenhui Ma, Bing Yang, Yongnian Dai, and Takayuki Watanabe. 2019. "Controllable synthesis of carbon nanomaterials by direct current arc discharge from the inner wall of the chamber." *Carbon* no. 142:278–284.

Zhou, Yue, Cun-Ku Dong, Li-Li Han, Jing Yang, and Xi-Wen Du. 2016. "Top-down preparation of active cobalt oxide catalyst." *ACS Catalysis* no. 6 (10):6699–6703.

Zhuang, Shiqiang, Eon Soo Lee, Lin Lei, Bharath Babu Nunna, Liyuan Kuang, and Wen Zhang. 2016. "Synthesis of nitrogen-doped graphene catalyst by high-energy wet ball milling for electrochemical systems." *International Journal of Energy Research* no. 40 (15):2136–2149.

2 Why SNM Play a Significant Role Today?

2.1 PROGRESS IN SNM

SNM are fundamentally distinct from regular materials due to their special, controllable, and functional properties. The majority of common SNM have fixed properties, which can change when new functional groups are added. These characteristics, however, are useful in SNM (Joshi and Bhattacharyya, 2011). The materials are capable of reacting to specific stimuli and eventually displaying new functional characteristics. The response of SNM is simple and immediate, in contrast to the complex common materials. SNM are divided into a number of subgroups based on how they react with different stimuli (Su and Song, 2021). The controlled application of external inputs, such as temperature, moisture, pH, electric field, and magnetic field, can significantly change the properties of smart materials. Different types of SNM are available based on their various properties, and their progress in different applications has been discussed below (Table 2.1).

SNM with controlled piezoelectricity can be used in a variety of industrial applications, especially those that involve vibrational generation and actuation. Following that, a wide range of commercially viable products were created based on the principle of piezoelectricity, including fuel injection, speakers, hydrophones, microphones, radio antenna oscillators, and timekeeping using quartz resonance (Topolov and Bowen, 2015). Following the advancement of nanotechnology, numerous piezoelectric SNM stimuli were created and can be produced as thin films, disks, or stacked sheets (Zaszczyńska et al., 2020).

Smart piezoelectric nanomaterials are being extensively researched for use in human health applications as nanotechnology develops (Thangudu 2020). The use of electrotherapy, a treatment method for neurological diseases based on electrical stimulation, dates back to the early 400 B.C. Torpedo fish are also frequently used to produce electric shocks that are used to lessen or control bodily pain. Since the knowledge on nanotechnology has advanced, it is now possible to store electricity in batteries. In the 1800s, piezoelectric materials were developed and electrical stimulation of tissues received more attention (Marino et al., 2017). Notably, even inside the cells, cells and tissues are highly responsive to applied electric fields. Therefore, this distinction is incredibly intriguing and heralds a new phase in the development of piezoelectric materials for medical applications. In order to stimulate cells and tissues, nanostructured piezoelectric interfaces are essential in nanomedicine. Properties of smart piezoelectric devices are drawing more attention than their roles as passive structural components or carriers for drugs. Additionally, stretchable, flexible, and reasonably priced wearable e-skins made of piezoelectric smart materials hold promise as tools for disease prevention,

DOI: 10.1201/9781003366270-2

TABLE 2.1
Different Types of SNM Available Based on Their Various Properties and Applications

Different Types of SNM	Properties	Applications
Piezoelectric nanomaterials	Ability to become electrically polarized when they are mechanically stimulated, and vice versa	Fuel injection, speakers, hydrophones, microphones, radio antenna oscillators, sensors
Electrochromic nanomaterials	Able to vary their coloration and transparency to solar radiation in a reversible manner, when they are subjected to a small electric field	Smart windows, displays, and anti-glare mirrors, smart textiles
Photochromic nanomaterials	Show a reversible change in optical properties (color) through the action of light, i.e., electromagnetic radiation	Optical switches, optical data storage devices, energy-conserving coatings, eye-protection glasses, and privacy shields
Magnetic nanomaterials	Response to an externally applied magnetic field	Several medical applications, such as cell isolation, immunoassay, diagnostic testing and drug delivery
Electro-responsive nanomaterials	Adjust their physical properties (size or shape) as response to a small change in the applied electric current	Drug delivery systems
Photo-responsive nanomaterials	Adjust their physical properties (size or shape) as response to a small change in the applied light	Biomedical and agrochemical applications
Thermally responsive nanomaterials	Able to go through a sudden change in their solubility as a reply to a small temperature change	Biomedical sciences and environmental remediation applications

monitoring one's health, tracking one's own physiological changes, and predicting early diseases (Trung and Lee, 2016) (Figure 2.1). In addition, thorough studies of piezoelectric materials on the body can be used to promote disease healing (Mahapatra et al., 2021).

Additionally, special features of electrochromic materials have been expanded in a variety of intriguing applications, such as active camouflages, smart windows, displays, and anti-glare mirrors (Figure 2.2). The use of electrochromic materials in smart windows is one possible application where they could reduce energy use and improve indoor comfort through reversible color changes. Due to their extraordinary qualities, photochromic materials (PC) are being further investigated for use in sensors and quick optical shutters. PC are widely used in ophthalmic sun screening applications and UV light protection eyewears in healthcare applications. Additionally, smart photochromic nanomaterials were used in cooling glasses; when exposed to sunlight, they darken and then turn colorless for user comfort and safety.

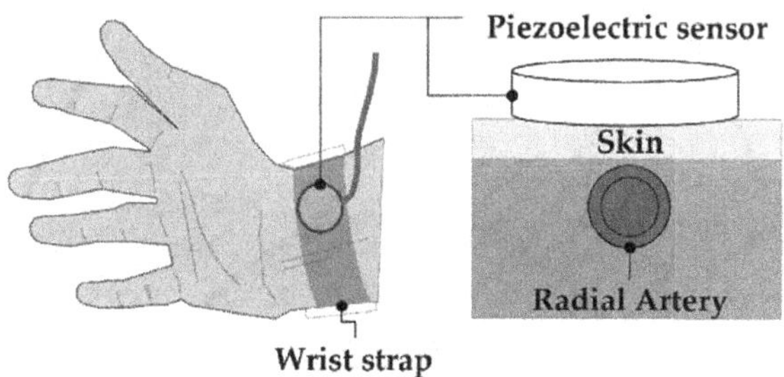

FIGURE 2.1 Piezoelectric sensor measurement location. (Wang and Lin, 2020.)

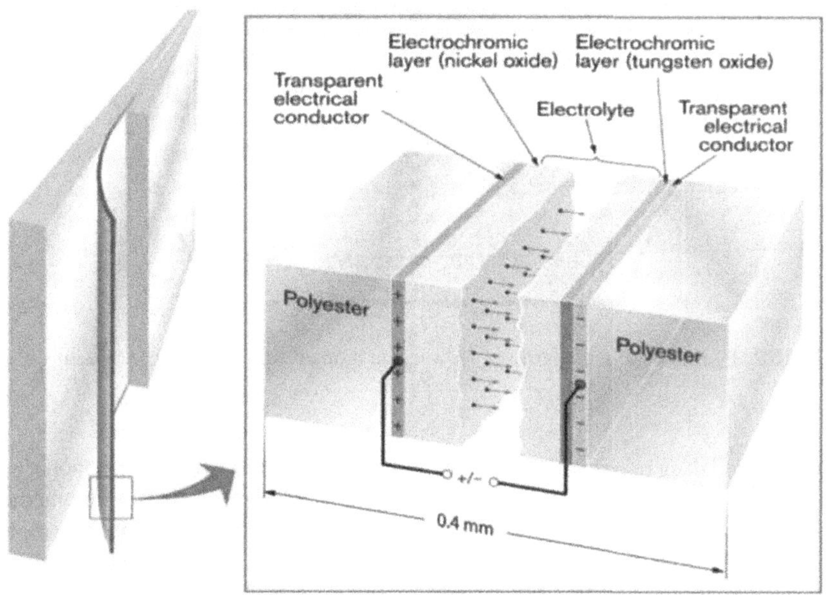

FIGURE 2.2 Principle design of a web-based EC device. Small arrows indicate ion transport upon application of a low voltage between the transparent electrical conductors. The foil can be used for glass lamination, as indicated in the left-hand part. (Granqvist et al., 2018.)

Magnetic nanomaterial development has proven to have a number of highly advantageous advantages in a wide range of industrial and commercial applications, including photonic, electronic, magnetic storage, and biomedical theranostics. "Magnetoresponsive materials" are defined as "materials that can respond to an applied magnetic field as a stimuli agent" (Eslami et al., 2022). Due to their special intrinsic chemical and physical properties, magnetic nanoparticles (MNPs) in particular are an emerging platform and receiving increased attention for their magnetic responsive-based applications in the biomedical field, such as magnetic hyperthermia, magnetic resonance imaging (MRI), and magnetic guided drug delivery (Figure 2.3).

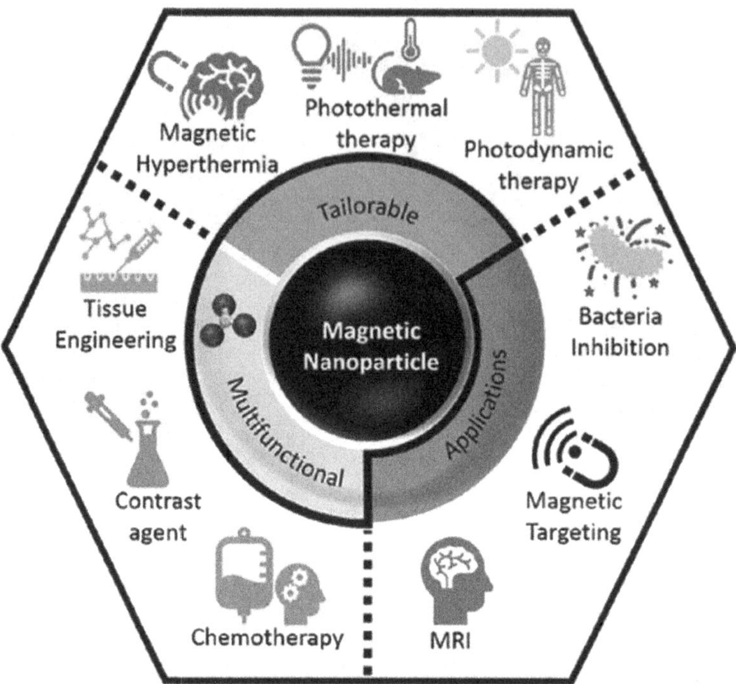

FIGURE 2.3 Magnetic nanoparticles for various biomedical applications. (Martins et al., 2021.)

Controlled drug release and drug accumulation on electro-responsive nanoparticle drug carriers like polymers and microgels are particularly intriguing in biomedical applications (Das et al., 2020). For instance, poly(ethylene imine) with a ferrocene end-group and encapsulated pyrene were combined to synthesize electro-responsive materials (Zhao et al., 2016). The release of the drug from its encapsulation causes the oxidation of ferrocene and further hydrophobic to hydrophilic nature transitions. A potential candidate for an electro-controlled release of drugs to treat various diseases should have redox properties.

SNM that can react to external light are known as photo-responsive materials. To achieve effective therapeutics, photo triggered SNM' controlled properties and non-intrusiveness are particularly crucial. Mechanisms used for photo triggered cargo delivery have in common that materials absorb electromagnetic radiation and convert it to various forms of energy (Figure 2.4).

The disruption of non-covalent interactions between drugs and particles by the photothermal effect has been widely used to release drug cargo. For instance, the hydrophilic-hydrophobic balance of thermally responsive polymers linked to plasmonic nanoparticles can be altered by the photothermal effect to trigger the release of drugs. Most of these polymers go through a volume phase transition at their lower critical solution temperature (LCST), above which they become insoluble. Below the LCST, the polymers are miscible in aqueous solutions. When thermo-responsive

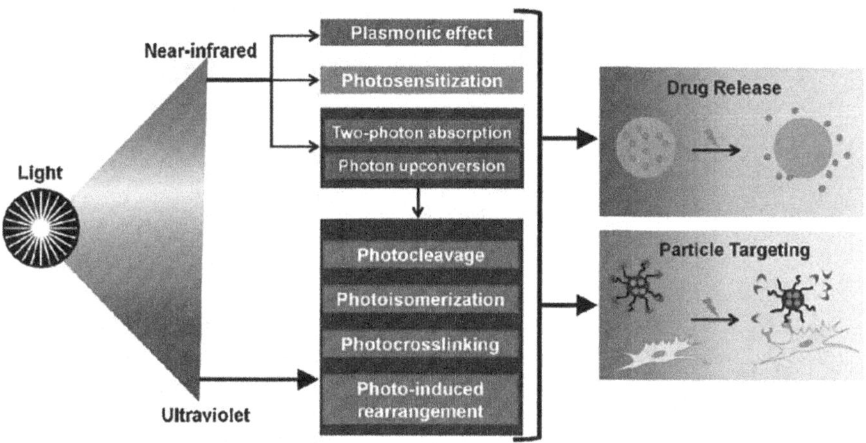

FIGURE 2.4 Mechanisms of photo responsiveness for nanoparticle targeting and drug release. (Rwei et al., 2015.)

polymers are included in nanoparticle systems, a temperature increase above the glass transition temperature (LCST) causes the polymers to become hydrophobic and causes them to collapse, resulting in the release of drugs that have been encapsulated.

Gold nanocages that have been covalently functionalized with the thermo responsive polymer poly(*N*-isopropylacrylamideco-acrylicamide) (pNIPAAm-co-pAAm) with an LCST of 39 °C are one illustration of such a system. The polymers were in an extended conformation at body temperature (37 °C), which sealed the pores in the nanocages and prevented dye release. The polymers disintegrated at temperatures higher than LCST, allowing dyes to escape through the pores (Yavuz et al., 2009).

Moreover, there is tremendous progress in the food packaging industry and smart textile industry with these SNM. Carbohydrate- and proteins-based nanoparticles are used in food packaging to enhance the strength and barrier properties such as water barrier properties (Zubair and Ullah, 2020). Carbon nanotubes are used to improve the mechanical and antimicrobial properties of the polymers in food packaging (Rezić et al., 2017). Carbon nanotubes are also used in oxygen sensors to monitor the oxygen concentration in modified atmospheric packaging. Carbon nanotubes are also incorporated into polymer matrices in food packaging to provide antimicrobial properties and intelligent sensors to detect the food spoilage (Ashfaq et al., 2022). More significantly, stimuli-responsive nanomaterials are widely used in food packaging as well (Figure 2.5) (Shafiq et al., 2020).

2.2 SAFE AND SUSTAINABLE INNOVATIONS OF SNM

In order to maximize functional and economic performance while minimizing negative effects on the environment and human health, sustainable nanotechnology is defined as "green chemistry applied to nanotechnology" (Gilbertson et al., 2015). Green chemistry principles can be used to create safer and more sustainable

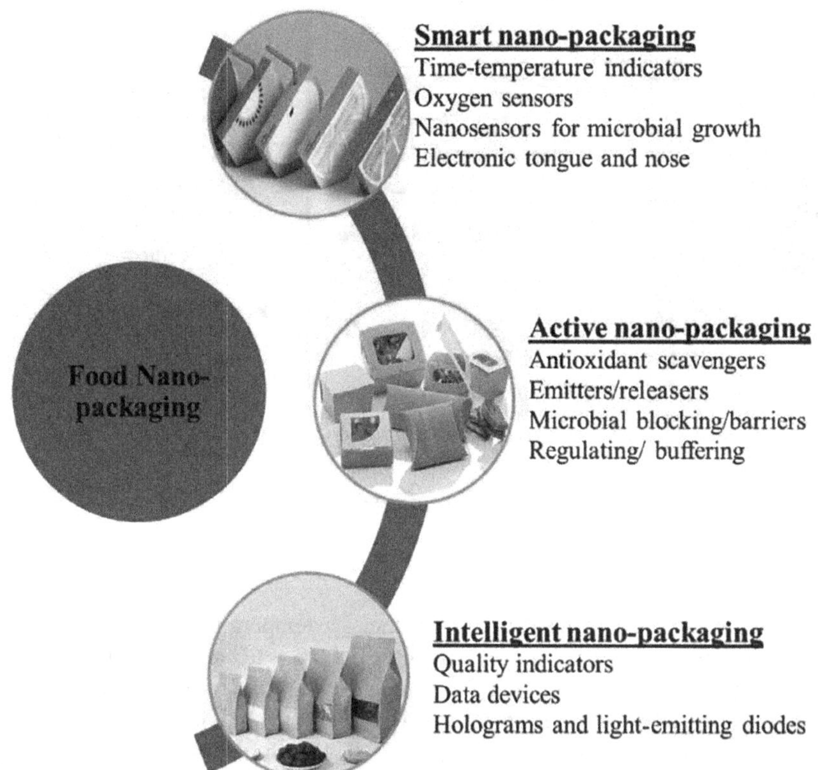

FIGURE 2.5 Classification of food nanopackaging. (Shafiq et al. 2020.)

nanomaterials as well as more effective and sustainable nano-manufacturing techniques. At the same time, nanotechnology is important for green innovation and green growth and is seen as a way to produce nanomaterials and products that are more sustainable or help to solve economic, environmental, and societal problems like the lack of energy or water filtration (Mauter et al., 2018). However, more investigation and analysis are required to ensure the development of green and smart nanotechnology in a responsible manner. This is because the costs of energy, waste, and resource extraction associated with the production of the materials used in these applications are still unknown (Gottardo et al., 2021).

Understanding the relationship between the nanomaterial's properties, its functional performance, and any potential negative effects on human health and the environment are essential for increasing societal acceptance of SNM (Falinski et al., 2018). There are not sufficient research information available about the potential toxicological effects on human health and the environment that may result from their use, despite the growing interest in SNM (Subramanian et al., 2010).

In controlled release technology and drug delivery systems, SNM are used as nanoencapsulates and nanocarriers (Vega-Vásquez et al., 2020). The COVID-19 crisis in particular has increased the need for efficient diagnostic and therapeutic tools to aid in the pandemic fight, and the stimuli-responsive properties of SNM are being investigated for the design of specialized controlled drug delivery systems, better antigen presentation, and immune modulation (Chauhan et al., 2020). The development of new and potent surface coatings that are resistant to viral adhesion may be aided by advancements in the field of biodegradable nanoparticles (Cagno et al., 2018).

SNM may aid in diagnosis as well as the treatment and prevention of diseases. Some approaches, such as the use of smart bio-sensing materials for diagnosis at point-of-care applications, are being investigated for nano-enabled bio-sensing systems as next-generation non-invasive disease diagnostics tools (Mujawar et al., 2020).

Controlled delivery systems have made progress in the last 10 years from pharmaceuticals to cosmetics and food, and they are now being considered for use in agriculture (Sundari and Anushree, 2017, Gottardo et al., 2021). For instance, a new class of stimuli-responsive nanocapsules that can treat various skin conditions, such as contact dermatitis and skin photodamage, has recently been developed and put in the market (Aziz et al., 2019). In response to pH changes or the presence of particular enzymes brought on by the skin condition that needs to be treated, the 'PeptiCaps' nanocapsules release the active ingredient in the desired location. Similar to this, some so-called "deodorants on request" use nanoencapsulates to release a scent or a bacteriophage-growth inhibitor in response to the pH, temperature, or moisture of the skin (Hofmeister et al., 2014).

Nanoencapsulates have been used in the food industry for a variety of purposes, including preventing the breakdown of nutrients, antioxidant molecules like ferulic acid and tocopherol, hydrophobic flavoring agents, or natural antimicrobial ingredients in the body, and allowing a slow release at a target specific location or upon the presence of a specific molecule (Piran et al., 2017).

In order to increase crop productivity and preserve soil integrity and function, Lowry et al. (2019) and Kah et al. (2019) recently reviewed the potential for creating smart nanostructures that allow for targeted (in terms of time, location, dose, and form) release and delivery of water, nutrients, agrochemicals, and antimicrobials to crops (Lowry et al., 2019, Kah et al., 2019). In order to create intelligent plant sensors for the real-time detection and monitoring of plant pathogens and stress conditions, nanotechnology can also be used in agriculture (Giraldo et al., 2019).

Food packaging that responds to environmental or food stimuli allows for real-time monitoring or correction of food quality and safety issues. The monitoring of gaseous substances like O_2, CO_2, or ethylene present in the headspace of the package is the most common application of stimuli-responsive nanomaterials in food packaging (Brockgreitens and Abbas, 2016). Due to their changes in shape, such as shrinking, swelling, or self-assembly, stimulus responsive nanomaterials are also suggested for use in corrective responsive food packaging. These changes can cause the release of a compound to stop things like microbial growth, the development of off flavors, off odors, color changes, or nutritional losses (Ghosh et al., 2019).

According to Brockgreitens and Abbas (2016), responsive packaging will have a significant impact on the food industry by lowering food waste, spoilage, recalls, and outbreaks of foodborne illness. This will promote sustainable development and safety (Brockgreitens and Abbas, 2016, Gottardo et al., 2021).

REFERENCES

Ashfaq, Alweera, Nazia Khursheed, Samra Fatima, Zayeema Anjum, and Kaiser Younis. 2022. "Application of nanotechnology in food packaging: Pros and Cons." *Journal of Agriculture and Food Research* no. 7:100270.

Aziz, Zarith Asyikin Abdul, Hasmida Mohd-Nasir, Akil Ahmad, Siti Hamidah Mohd. Setapar, Wong Lee Peng, Sing Chuong Chuo, Asma Khatoon, Khalid Umar, Asim Ali Yaqoob, and Mohamad Nasir Mohamad Ibrahim. 2019. "Role of nanotechnology for design and development of cosmeceutical: Application in makeup and skin care." *Frontiers in Chemistry* no. 7:739.

Brockgreitens, John, and Abdennour Abbas. 2016. "Responsive food packaging: Recent progress and technological prospects." *Comprehensive Reviews in Food Science and Food Safety* no. 15 (1):3–15.

Cagno, Valeria, Patrizia Andreozzi, Marco D'Alicarnasso, Paulo Jacob Silva, Marie Mueller, Marie Galloux, Ronan Le Goffic, Samuel T Jones, Marta Vallino, and Jan Hodek. 2018. "Broad-spectrum non-toxic antiviral nanoparticles with a virucidal inhibition mechanism." *Nature Materials* no. 17 (2):195–203.

Chauhan, Gaurav, Marc J Madou, Sourav Kalra, Vianni Chopra, Deepa Ghosh, and Sergio O Martinez-Chapa. 2020. "Nanotechnology for COVID-19: Therapeutics and vaccine research." *ACS Nano* no. 14 (7):7760–7782.

Das, Sabya Sachi, Priyanshu Bharadwaj, Muhammad Bilal, Mahmood Barani, Abbas Rahdar, Pablo Taboada, Simona Bungau, and George Z Kyzas. 2020. "Stimuli-responsive polymeric nanocarriers for drug delivery, imaging, and theragnosis." *Polymers* no. 12 (6):1397.

Eslami, Parisa, Martin Albino, Francesca Scavone, Federica Chiellini, Andrea Morelli, Giovanni Baldi, Laura Cappiello, Saer Doumett, Giada Lorenzi, and Costanza Ravagli. 2022. "Smart magnetic nanocarriers for multi-stimuli on-demand drug delivery." *Nanomaterials* no. 12 (3):303.

Falinski, Mark M, Desiree L Plata, Shauhrat S Chopra, Thomas L Theis, Leanne M Gilbertson, and Julie B Zimmerman. 2018. "A framework for sustainable nanomaterial selection and design based on performance, hazard, and economic considerations." *Nature Nanotechnology* no. 13 (8):708–714.

Ghosh, Chandan, Debabrata Bera, and Lakshmishri Roy. 2019. "Role of nanomaterials in food preservation." In Ram Prasad (Ed.), *Microbial Nanobionics*, 181–211. Springer, Berlin/Heidelberg.

Gilbertson, Leanne M, Julie B Zimmerman, Desiree L Plata, James E Hutchison, and Paul T Anastas. 2015. "Designing nanomaterials to maximize performance and minimize undesirable implications guided by the Principles of Green Chemistry." *Chemical Society Reviews* no. 44 (16):5758–5777.

Giraldo, Juan Pablo, Honghong Wu, Gregory Michael Newkirk, and Sebastian Kruss. 2019. "Nanobiotechnology approaches for engineering smart plant sensors." *Nature Nanotechnology* no. 14 (6):541–553.

Gottardo, Stefania, Agnieszka Mech, Jana Drbohlavová, Aleksandra Małyska, Søren Bøwadt, Juan Riego Sintes, and Hubert Rauscher. 2021. "Towards safe and sustainable innovation in nanotechnology: State-of-play for smart nanomaterials." *NanoImpact* no. 21:100297.

Granqvist, Claes G, Ilknur Bayrak Pehlivan, and Gunnar A Niklasson. 2018. "Electrochromics on a roll: Web-coating and lamination for smart windows." *Surface and Coatings Technology* no. 336:133–138.

Hofmeister, Ines, Katharina Landfester, and Andreas Taden. 2014. "pH-sensitive nanocapsules with barrier properties: Fragrance encapsulation and controlled release." *Macromolecules* no. 47 (16):5768–5773.

Joshi, Mangala, and Anita Bhattacharyya. 2011. "Nanotechnology–a new route to high-performance functional textiles." *Textile Progress* no. 43 (3):155–233.

Kah, Melanie, Nathalie Tufenkji, and Jason C White. 2019. "Nano-enabled strategies to enhance crop nutrition and protection." *Nature Nanotechnology* no. 14 (6):532–540.

Lowry, Gregory V, Astrid Avellan, and Leanne M Gilbertson. 2019. "Opportunities and challenges for nanotechnology in the agri-tech revolution." *Nature Nanotechnology* no. 14 (6):517–522.

Mahapatra, Susmriti Das, Preetam Chandan Mohapatra, Adrianus Indrat Aria, Graham Christie, Yogendra Kumar Mishra, Stephan Hofmann, and Vijay Kumar Thakur. 2021. "Piezoelectric materials for energy harvesting and sensing applications: Roadmap for future smart materials." *Advanced Science* no. 8 (17):2100864.

Marino, Attilio, Giada Graziana Genchi, Edoardo Sinibaldi, and Gianni Ciofani. 2017. "Piezoelectric effects of materials on bio-interfaces." *ACS Applied Materials & Interfaces* no. 9 (21):17663–17680.

Martins, Pedro M, Ana C Lima, Sylvie Ribeiro, Senentxu Lanceros-Mendez, and Pedro Martins. 2021. "Magnetic nanoparticles for biomedical applications: From the soul of the earth to the deep history of ourselves." *ACS Applied Bio Materials* no. 4 (8):5839–5870.

Mauter, Meagan S, Ines Zucker, Francois Perreault, Jay R Werber, Jae-Hong Kim, and Menachem Elimelech. 2018. "The role of nanotechnology in tackling global water challenges." *Nature Sustainability* no. 1 (4):166–175.

Mujawar, Mubarak A, Hardik Gohel, Sheetal Kaushik Bhardwaj, Sesha Srinivasan, Nicolerta Hickman, and Ajeet Kaushik. 2020. "Nano-enabled biosensing systems for intelligent healthcare: Towards COVID-19 management." *Materials Today Chemistry* no. 17:100306.

Piran, Parizad, Hossein Samadi Kafil, Saeed Ghanbarzadeh, Rezvan Safdari, and Hamed Hamishehkar. 2017. "Formulation of menthol-loaded nanostructured lipid carriers to enhance its antimicrobial activity for food preservation." *Advanced Pharmaceutical Bulletin* no. 7 (2):261.

Rezić, Iva, Tatjana Haramina, and Tonči Rezić. 2017. "Metal nanoparticles and carbon nanotubes—perfect antimicrobial nano-fillers in polymer-based food packaging materials." In Alexandru Mihai Grumezescu (Ed.), *Food packaging*, 497–532. Elsevier, Cambridge, MA.

Rwei, Alina Y, Weiping Wang, and Daniel S Kohane. 2015. "Photoresponsive nanoparticles for drug delivery." *Nano today* no. 10 (4):451–467.

Shafiq, Mehwish, Sumaira Anjum, Christophe Hano, Iram Anjum, and Bilal Haider Abbasi. 2020. "An overview of the applications of nanomaterials and nanodevices in the food industry." *Foods* no. 9 (2):148.

Su, Meng, and Yanlin Song. 2021. "Printable smart materials and devices: Strategies and applications." *Chemical Reviews* no. 122 (5):5144–5164.

Subramanian, Vrishali, Jan Youtie, Alan L Porter, and Philip Shapira. 2010. "Is there a shift to "active nanostructures"?" *Journal of Nanoparticle Research* no. 12 (1):1–10.

Sundari, P Tripura, and H Anushree. 2017. "Novel delivery systems: Current trend in cosmetic industry." *European Journal of Pharmaceutical and Medical Research* no. 4 (8):617–627.

Thangudu, Suresh. 2020. "Next generation nanomaterials: Smart nanomaterials, significance, and biomedical applications." In Firdos Alam Khan (Ed.), *Applications of Nanomaterials in Human Health*, 287–312. Springer, Berlin/Heidelberg.

Topolov, V Yu, and Chris R Bowen. 2015. "High-performance 1–3-type lead-free piezocomposites with auxetic polyethylene matrices." *Materials Letters* no. 142:265–268.

Trung, Tran Quang, and Nae-Eung Lee. 2016. "Flexible and stretchable physical sensor integrated platforms for wearable human-activity monitoring and personal healthcare." *Advanced Materials* no. 28 (22):4338–4372.

Vega-Vásquez, Pablo, Nathan S Mosier, and Joseph Irudayaraj. 2020. "Nanoscale drug delivery systems: From medicine to agriculture." *Frontiers in Bioengineering and Biotechnology* no. 8:79.

Wang, Ting-Wei, and Shien-Fong Lin. 2020. "Wearable piezoelectric-based system for continuous beat-to-beat blood pressure measurement." *Sensors* no. 20 (3):851.

Yavuz, Mustafa S, Yiyun Cheng, Jingyi Chen, Claire M Cobley, Qiang Zhang, Matthew Rycenga, Jingwei Xie, Chulhong Kim, Kwang H Song, and Andrea G Schwartz. 2009. "Gold nanocages covered by smart polymers for controlled release with near-infrared light." *Nature Materials* no. 8 (12):935–939.

Zaszczyńska, Angelika, Arkadiusz Gradys, and Paweł Sajkiewicz. 2020. "Progress in the applications of smart piezoelectric materials for medical devices." *Polymers* no. 12 (11):2754.

Zhao, Yi, Ana C Tavares, and Marc A Gauthier. 2016. "Nano-engineered electro-responsive drug delivery systems." *Journal of Materials Chemistry B* no. 4 (18):3019–3030.

Zubair, Muhammad, and Aman Ullah. 2020. "Recent advances in protein derived bionanocomposites for food packaging applications." *Critical Reviews in Food Science and Nutrition* no. 60 (3):406–434.

3 SNM for Biomedical Applications

3.1 NANOPARTICLE BASED BIOSENSORS

Research on biologically recognizable elements with specific functions has enabled the development of a new field of focus on electrochemically modified electrodes called biosensors. Nowadays, there is considerable interest in reliable and affordable sensors and detection systems for various applications. Recent concerns about environmental exposure to both biological and chemical agents have been critical to the development of new sensor and detector technologies. New materials are being developed to meet these challenges ahead and SNM appear to be one of the key solutions for these challenges (Li et al., 2012). For example, implantable biosensors have been developed using SNM which are capable of providing continuous data on the levels of a targeted analyte (Figure 3.1). This has enabled the trends and changes in the analyte levels over time to be monitored without any need for intervention from either a patient or a clinician (Gray et al., 2018).

A biosensor is broadly defined as "a self-contained analyzer containing a biologically active substance which is in close contact with the appropriate transforming element to detect the concentration or activity of a chemical species in any type of sample" (Karunakaran et al., 2015). Biosensors based on nanomaterials are currently studied in medicine to detect changes at the individual molecule, individual cell, or molecular level (Choi et al., 2014). The ideal biosensor should have the following characteristics:

1. It is necessary to show high specific performance and excellent stability under various environmental conditions.
2. The response should be clear, accurate, reproducible, and linear throughout the analytical range without interfering with the electrical signal.
3. Feedback should be independent of external factors such as pH, temperature, ionic, and chemical concentrations.

When using biosensors in clinical settings, the probe should be biocompatible, nontoxic, and should not elicit an immune response. Biosensors can also be subdivided based on bioreceptors, transducers, detection systems, and the technology (Figure 3.2).

Based on the biorecognition concept, biosensors can be catalytic or non-catalytic. In a catalytic biosensor, the interaction of the analyte and the bioreceptor produce a particular biochemical reaction product. This kind of biosensor can detect enzymes, bacteria, tissues, whole cells, etc. In a non-catalytic biosensor, the analyte and receptor are irreversibly coupled, and no new biochemical reaction product is produced as

DOI: 10.1201/9781003366270-3

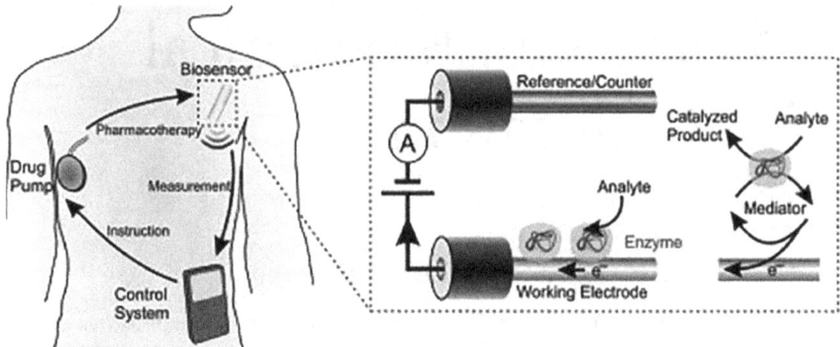

FIGURE 3.1 Schematic illustration of implantable biosensor in human heart. (Scholten and Meng, 2018.)

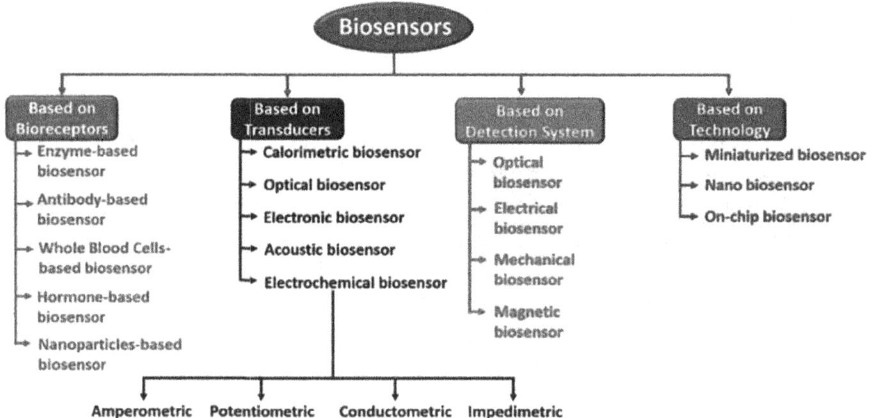

FIGURE 3.2 Classification of biosensors depending on the diverse use of bioreceptors and transducers. (Kulkarni et al., 2022.)

a result of the interaction. These sensors interact with biological components such as nucleic acid, antibodies, and cell receptors to recognize the target.

Enzyme-based chemical biosensors are based on biological recognition. In order to operate, the enzymes must be available to catalyze a specific biochemical reaction and be stable under the normal operating conditions of the biosensor (Rocchitta et al., 2016). Antibody-based sensors can be attached to antibodies or ligands that may influence antibody–antigen interactions (North, 1985). Hormones are typically produced by glands or specific cell types and released into the bloodstream where they target cells. Human disease diagnosis and treatment methods are now effectively using hormone electrochemical biosensing.

Nanotechnology advancements have broadened the scope of biosensor research. Exploring nanomaterials such as carbon nanotubes, quantum dots, carbon nanoparticles, various nanocomposites, nanorods, nanowires, nanoflakes, nanocones, metal,

metal oxide-based nanoparticles, etc., open up the possibility to improve the biosensor performance and thereby increase the detection power by adjusting its size and shape (Zhang et al., 2017). Nanoparticles-based biosensors operate under the same principle as their macro- and micro-counterparts but signals and data are translated at the nanoscale. Nano-biosensors are used in (1) monitoring of physical and chemical events, (2) detecting of biochemicals in various cellular organelles, (3) detecting nanoparticles in environmental and industrial applications, and (4) detecting of contaminants that may be dangerous at extremely low concentrations.

Inorganic nanoparticles are usually promising materials in fabrication of nanosensors with unique electronic and electrode catalytic properties and those properties mainly depend on the particle size and the morphology of nanoparticles. Among the properties of nanoparticles, the efficiency of electronic and electrochemical redox reactions is important and it makes these nanoparticles of particular interest suitable for biomedical applications. Gold nanoparticles (AuNPs) have been widely studied in biomedical applications due to its unique adjustable optical properties. The AuNPs are extensively used as an imaging, detection, and sensing tool. In addition, AuNPs are attractive candidates for photothermal therapy, diagnostics, and drug delivery systems. Further, its ability to amplify electronic signals helps in biosensing when the surface of the Au nanostructure contacts the biological components. The use of AuNPs also provides a new approach to the development of enzyme-based biosensors. For example, AuNPs are used to amplify both Rayleigh and Raman signals to obtain chemical information in cancer cells (Huang and El-Sayed, 2010). It is also used to detect biological molecules and cells according to the biological environment due to its sensitivity and selectivity. Various formulations of AuNPs have been developed to target biological molecules/components such as DNA, RNA, cells, organic compounds, and proteins (Khalil et al., 2016).

Transducers transform information into quantitative displays such as electrical signals. Transduction can use the five different detection methods: (1) electrochemical, (2) optical, (3) electronic, (4) acoustic, and (5) calorimetric (Kulkarni et al., 2022). Electrochemical detection usually uses fluorescence characteristics and detection methodologies (Grieshaber et al., 2008). The electrochemical biosensor can detect specific molecules having high specificity based on the relationship between the transformation element and its biological aspect. Optical detection comprises various types of spectroscopy techniques such as absorption, fluorescence, Raman, and large-scale analysis involves the use of "piezoelectric crystals." These crystals can vibrate at a specific frequency by applying an electrical signal within the targeted range. Acoustic detection is functioned by altering an acoustic wave's physical characteristics in response to an alteration in the amount of analyte absorbed (Fogel et al., 2016). Calorimetric biosensors measure the change in temperature of the solution containing the analyte following enzyme action and interpret it in terms of the analyte concentration in the solutions (Kulkarni et al., 2022).

3.1.1 Advantages of Using Nanomaterials for Biosensors

In order to fabricate miniature biosensor devices, nanomaterials are being investigated for use in a variety of fields, including physics, medicine, biomedicine, and chemistry. Nanomaterials are expected to have a significant impact on biosensors in medical applications due to their similar size compared with biological materials such as enzymes, nucleotides, proteins, and antibodies. In biomedical engineering, from drug delivery to biosensors, the remarkable properties of nanomaterials, such as high surface area to volume ratio, optical emission tuning properties, electrical, and magnetic properties, are widely utilized (Pandey et al., 2008).

The nanostructured metal NPs (Au, Ag, Ni, etc.), semiconductors (CeO_2, ZiO, SnO_2, TiO_2, etc.), and multiwalled carbon nanotubes are widely used in the development of biosensor devices due to their unique and sophisticated characteristics. Furthermore, these nanoparticles can be used to enhance the immobilization of the biomolecules due to its intrinsic properties and large surface area. As an example, indium (III) oxide (In_2O_3) nanowire type biosensors are used in electrical detection of the severe acute respiratory syndrome virus N-protein (Ishikawa et al., 2009). Luo et al. have fabricated a boron-doped diamond nanorod type electrode for the detection of non-enzymatic glucose via an amperometry technique (Luo et al., 2009). Law et al. have also reported a NP-based biosensor to detect antigen concentration in femtomolar by integrating both the NPs and immunoassay sensing technologies into a phase interrogation surface plasmon resonance (SPR) system (Law et al., 2011). Choi et al. have prepared flexible conductive reduced graphene oxide/Nafion (RGON) hybrid films using solution chemistry that utilizes self-assembly and directional convective-assembly (Choi et al., 2010). Therefore, nanomaterials can be applied for immobilization of various biomolecules, for amplification of several signals, and as mediators, electroactive species, and detection nanoprobes. Different nanomaterials are being utilized in biosensors based on the transduction mechanism such as electrochemical, optical, thermoelectric, etc. (Malhotra and Ali, 2018).

3.1.2 Biosensing of DNA

DNA biosensing is an area where nanotechnology has the potential to make significant advances in medicine. In recent years, DNA detection has attracted more attention for its applications in medical research, drug delivery and diagnosis and there are different types of DNA biosensors (Figure 3.3). Functional DNA strand-based biosensors, such as aptamer and DNAzyme biosensors, are biosensors that recognize specific targets by using functional DNA strands. Biosensors based on the DNA hairpin, the hybridization chain reaction (HCR), and catalytic hairpin assembly (CHA) are examples of DNA hybridization-based biosensors that are used in enzyme-free nucleic acid amplification techniques to improve their responses. As with biosensors based on the DNA tetrahedron and DNA origami, DNA template-based biosensors refer to biosensors that are embellished with a DNA template (the super molecule DNA assembly structures with programmable anchoring points) (Loretan et al., 2020).

Recently, methods have been developed to detect DNA, aptamers, and oligonucleotides using chemically functionalized AuNPs. DNA detection is essential for photogene detection and biomedical research. The biosensor works by introducing the DNA strand containing the analyte into the solution (Abu-Salah et al., 2015). The analyte hybridizes to the test DNA to create a double strand. Because the test

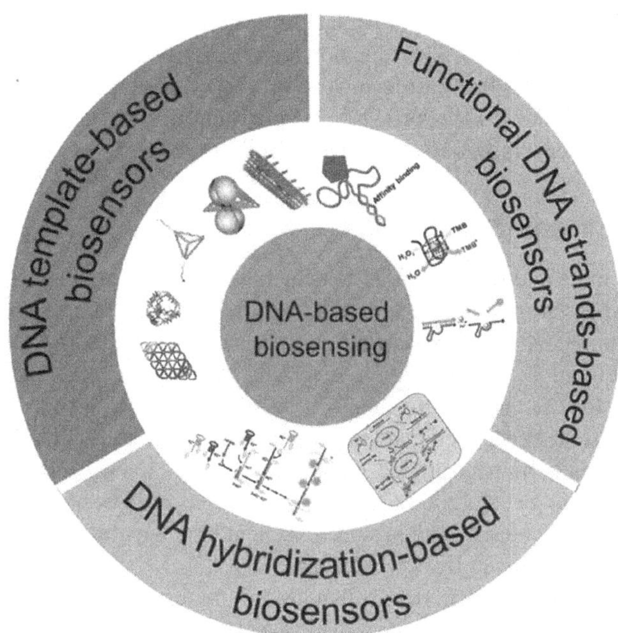

FIGURE 3.3 Schematic diagram of different DNA-based biosensors. (Hua et al., 2022.)

sequence is short and weighs only one base pair, the sequence is too small to be measured by traditional quantitative or qualitative methods. However, biosensors make it possible to record, amplify, and measure the responses. In other words, a notable challenge for DNA biosensors is to amplify the effects of hybridization.

Many DNA sensor systems contain AuNPs that attach to DNA, increasing the detection limit and sensitivity of the assay. Auyeung et al. (2014) have used a combination of quantum optics effects and molecular recognition. This research group proposed a method for assembling colloidal nanodots that attach to the surface of two batches of 13 nm gold particles with non-complementary DNA oligonucleotides capped with thiol groups that bind DNA to AuNPs. The biosensor detects the target DNA strands and changes the composition of the gold nanospheres as the strands rearrange and approach each other. This type of sensor is known as a "color measurement sensor." This approach uses common detection principles and methods used to search for small pieces of DNA that are amplified and recorded by the transducer. Duyne et al. used AuNPs lithography to create small gold dots on the surfaces. Nanostructures with biological-binding sites on AuNPs make it easy to detect protein analysts such as antibodies that bind to biological invaders (antigens) in the body (Sagle et al., 2012). This group demonstrated that nanoparticles can measure a single molecule in a particular analyte. Elghani et al. have reported the use of AuNPs modified with mercaptoalkyl oligonucleotide molecules and developed a colorimetric detection method for polynucleotide molecules (Elghanian et al., 1997). The unique property of AuNPs to induce color changes and the versatility of graphene oxides (GO) in surface modification makes them ideal in the application of colorimetric biosensor. In this regard, a label free optical method has been developed by Huang et al to detect DNA hybridization through a visually observed color change (Thavanathan et al., 2014). Apart from AuNPs, carbon nanotubes, peróvskites, copper nanoparticles, nickel nanoparticles, and platinum nanoparticles have also been used in DNA biosensors (Jeong et al., 2013, Yin et al., 2014, Dhara et al., 2014).

3.1.3 Detection of Cells

Fluorescence microscopy can be used to detect tumor cells, and it has few disadvantages. The disadvantages include low sensitivity and tedious sample preparation for imaging. A functionalized nanoparticle sensing system can be used for biomedical diagnostics (Figure 3.2). In addition to detecting biomolecules, AuNPs have proven advantageous for the targeted diagnosis of cancer biomarkers and the detection of cells such as breast cancer cells and oral epithelial cancer cells. The preparation of nanocomposite gels by neutralizing a solution of AuNPs encapsulated in chitosan was reported by Ding et al. (2007). Nanocomposite gels have been shown to improve cell immobilization capacity and provide excellent biocompatibility for maintaining the activity of immobilized live cells (Figure 3.4).

Sensitive detection of cancerous cells plays an important role in the early detection of cancer and cancer metastasis. However, since circulating tumor cells are extremely rare in peripheral blood, the detection of cancer cells with high sensitivity and specificity remains challenging. In order to determine whether a tumor is present, whether it is cancerous or benign, and whether treatment has been successful in reducing or eliminating cancerous cells, biosensors measure the levels of specific proteins expressed and/or secreted by tumor cells. Given that most cancers have multiple biomarkers, biosensors that can detect multiple analytes may be especially

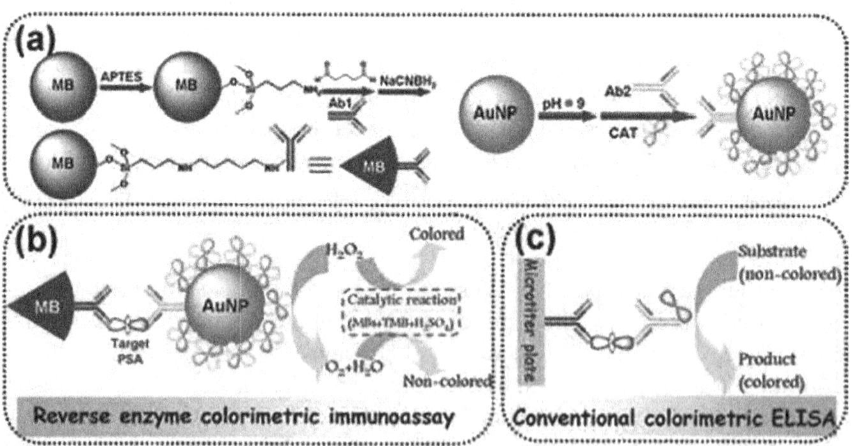

FIGURE 3.4 Magneto-controlled enzyme-mediated reverse colorimetric immunoassay protocol: (a) design of monoclonal mouse anti-human PSA-conjugated magnetic beads (MB-Ab1) and polyclonal goat anti-human PSA/catalase-labeled AuNP (multi-CATAuNP-Ab2), (b) Magneto-controlled enzyme-mediated reverse colorimetric immunosensing strategy, and (c) Conventional colorimetric enzyme-linked immunosorbent assay (ELISA). (Gao et al., 2013.)

helpful in the diagnosis and monitoring of cancer. A biosensor's capacity to simultaneously test for several markers not only aids in diagnosis but also saves time and money (Bohunicky and Mousa, 2010). As an example, Zhou and co-workers have developed a method which is simple, sensitive and specific detection of cancer cells with the limit of detection sensitivity of four cancer cells HCR to multibranched HCR (mHCR). mHCR can produce long products with multiple biotins for signal amplification and multiple branched arms for multivalent binding (Zhou et al., 2014). An aptamer–nanoparticle strip biosensor (ANSB) has been developed by Liu et al for the rapid, specific, sensitive, and low-cost detection of circulating cancer cells (Liu et al., 2009).

3.1.4 Applications of Biochemical Biosensors

Biosensors have great potential to detect a wide range of analysts in many areas. The ultimate goal of development of biosensor is to improve speed, sensitivity, accuracy, and sample preparation. As many clinical diagnostic biomarker researchers have identified new proteins and chemical markers associated with pathological conditions, companies have also begun developing asymptomatic diagnostic tools that can detect known markers at asymptomatic levels (Sin et al., 2014). Common analysts for medical biosensing include DNA, glucose, lactic acid, creatinine, urea, cholesterol, and uric acid. Glucose is the most common target analyte because diabetes is a global health problem caused by changes in glucose metabolism. Glucose oxidase is used as a biomolecular sensing element in glucose biosensors, and biosensor electrochemical signals are generated by interacting with glucose. Glucose concentration is one of the main indicators of diseases such as diabetes and endocrine disorders. Recent research has focused on micro-type and nano-type biosensors for implanting glucose monitoring systems in the body. Ning et al. have reported a new carbon nanotube-gold-titanium nanocomposites based on glucose biosensors. They developed a biosensor by immobilizing glucose oxidase on a carbon nanotube-gold titanium oxide nanocomposite modified with a glassy carbon electrode (Ning et al. 2019). Phosphorescent detection of glucose in human serum was developed by Ho et al. (2014). They manufactured optical sensors based on sensor substrates made of crystalline iridium-containing coordination compounds. Creatinine biosensors made with potential difference measurements or amperometry tools can detect kidney, thyroid, and muscle disorders. Creatinine is produced by the biological system of creatine, phosphocreatine, and adenosine triphosphate and is filtered from the bloodstream by the kidneys. Meyerhoff and Rechnitz developed a potentiometric titration biosensor based on tripolyphosphate-activated creatinase in 1976. The sensor had a highly selective enzyme electrode for creatinine detection. Electrodes made using a new microbial enzyme (creatinine deminase) were immobilized on collagen and the intestinal membrane of pigs in combination with ammonia electrodes and were developed by Guilbault and Coulet. Huang and Shish describe a newly established potentiometric system. It is a highly sensitive creatinine biosensor obtained by cross-linking creatine deminase on a polyaniline–nafion composite electrode that effectively senses ammonium ions. These potentiometric systems are based on the catalysis of creatinine by creatinine iminohydrolase on the surface of

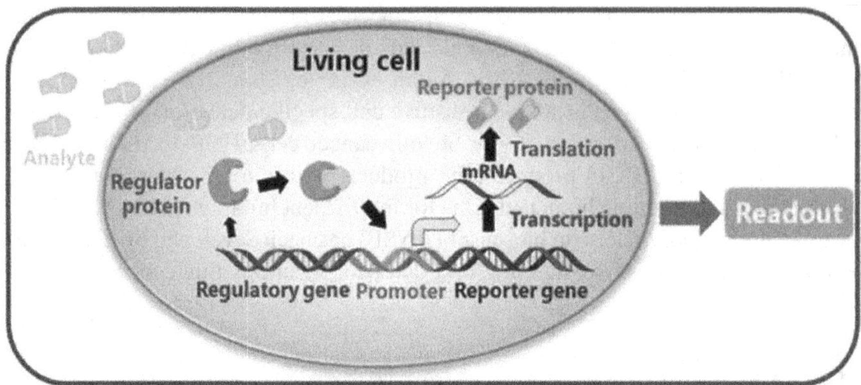

FIGURE 3.5 A schematic diagram representing a typical whole cell-based biosensor. (Gui et al., 2017.)

ion-selective ammonium detection electrodes. This is relatively simple as it requires only a single enzyme. These systems have significant drawbacks due to interference from the endogenous ammonium ions in the blood and urine. Therefore, the results cannot be used for confirmatory diagnostics. Urea is one of the end products of protein metabolism and is a very important analyte. It can be used to detect renal dysfunction and related diseases. Urease, an enzyme that catalyzes the hydrolysis of urea to produce ammonia and carbon dioxide, is commonly used as a bio detective element in urea sensor systems. An important step in the manufacture of enzyme-based biosensors is the immobilization of enzymes on the electrode surface. To date, many materials have been used to immobilize this enzyme, including polyvinyl alcohol (PVA), polyacrylamide membranes, poly (ethylene glycol) diglycidyl ethers (PEGDE), fibers, NiO nanoparticles, nanoporous alumina membranes, titanium, and silica. Covalent enzyme immobilization has been done with PEGDE in order to produce microelectrode biosensors which measure urease on poly (vinylferrosenium) films. Urea biosensors are developed by using biosensing layers (bio transducers) to detect changes in the physical environment and convert them into electrical signals to provide quantitative data. Conductive polymers are also widely used in the manufacture of bio transducers. The main advantage in bio converters is that they allow for the physicochemical changes required to convert a signal into an electrical signal. Early and accurate detection of various biochemical or biological disorders in the body using smart materials may be the key to preventing diseases such as cancer from progressing to advanced stages. Using biosensors for early whole cell detection may be a way to provide better medical care to patients, as treatment is almost impossible in advanced stages (Figure 3.5).

3.2 SMART NANOPARTICLES FOR PHOTODYNAMIC THERAPY

Photodynamic therapy (PDT) is used in a variety of clinical therapies that destroy diseased tissue or cells by combining photosensitizers with light in the presence of the appropriate amount of molecular oxygen at the target site. With minimal

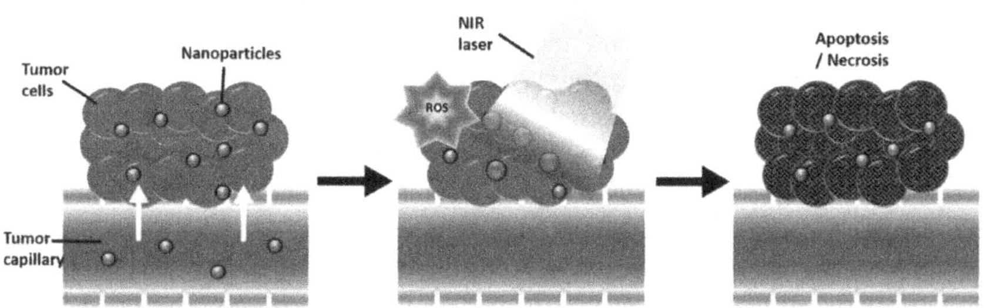

FIGURE 3.6 Schematic illustration of the physiological and biological effects of gold nanoparticle-mediated photothermal therapy (PTT) and PDT. A large amount of gold nanoparticles accumulate due to the leaky vasculature of the tumor, resulting in a photothermal effect in response to near-infrared (NIR) light and reactive oxygen species (ROS) generated by secondary delivered photosensitizer (PS), ultimately inducing apoptosis and necrosis of tumor tissue. (Kim and Lee 2018.)

invasiveness, selectivity, and topical therapeutic action, it can be applied to infectious diseases such as precancerous lesions, early cancer, and age-related macular degeneration (AMD). Mechanically, local or systemic delivery of PDT to tissue via nanoparticles involves three major steps: excitation, reactive oxygen species (ROS) formation, and finally cell death (Figure 3.6).

ROS reacts with basal molecular oxygen to produce superoxide anion radicals, hydroxyl radicals, and hydrogen peroxide (type I reaction). The formation of singlet oxygen by energy transfer from a photoactivated photosensitizer to molecular oxygen (type II reaction) is always part of the molecular process initiated by PDT. Sufficient formation of singlet oxygen leads to irreversible destruction of lesioned tissue without affecting the surrounding healthy tissue. Singlet oxygen has a lifespan of less than 3.5 seconds and can diffuse from 0.01 to 0.02 m. Therefore, the extent of damage is limited to areas with sufficient concentrations of photosensitizer drug molecules. In general, DNA damage is rare because the nucleus and nuclear envelope are protected. Photosensitizers can accumulate intracellularly in the plasma membrane, golgi apparatus, mitochondria, endoplasmic reticulum, endosomes, and lysosomes, so these structures can be damaged by photo-oxidative phototoxicity. Damage activates various signaling pathways involved in cell death via necrosis, apoptosis, and autophagy. Activation of the immune system during the process of PDT also occurs when the vascular system of the tissue is destroyed, causing local necrosis. Factors that play an important role in PDT are the type of photosensitizer, the concentration of the photosensitizer, whether the administration is extracellular or intracellular, the amount and wavelength of light (the selected light is the maximum absorption wavelength of the drug molecule). On the other hand, smart NPs provide a faster conversion to a singlet excited state, the time between photosensitizer administration and exposure, the production of singlet oxygen species, and the cell type or tissue that is finally processed. Most photosensitizers are hydrophobic in nature and tend to aggregate in aqueous systems. Aggregation leads to lower quantum yields and increases the complexity of intravenous administration. Previous studies have also shown the low photosensitizer selectivity, low wavelength absorption spectra, low excitation coefficients, and high accumulation in skin tissue. Several approaches have been developed to overcome these limitations and use PDT to target damage to cancerous tissue without affecting healthy tissues. Unique features of cancer cells such as oxidation status, specific cell surface antigens, and low-density lipoprotein can be used for PDT targeting. Coupling of anti-cell-specific ligands can be used to increase the effectiveness of PDT. Various carriers are used in PDT photosensitizers. Polymer micelles, liposomes, quantum dots, nanoparticles which (smaller than 1 μm in size) can improve the effectiveness of PDT. They provide a large surface area that can modify the surface of photosensitizers for a variety of purposes, such as improving biocompatibility and coupling specific ligands. NPs have a large distribution that is efficiently taken up by cells. NPs are also used in drug release control applications in various biomedical settings. Given these benefits of NPs, researchers have used different types of NPs to increase the effectiveness of PDT. Depending on how the photosensitizer NPs (PNPs) is activated, it is called active nanoparticles (active PNPs such as gold NPs) or passive nanoparticles (passive PNPs such as quantum dots and magnetic NPs).

The NPs used to activate the photosensitizer drug molecule must meet certain important criteria.

- The NPs emission band must coincide with the photosensitizer absorption band, which ensures efficient activation of the photosensitizer and formation of singlet oxygen species.
- The NPs must be compatible with surface modifications that allow the attachment of photosensitizers.
- The NPs must have high luminous efficiency so that the absorbed energy can be used efficiently without delaying the reaction.
- NPs need to be more photostable than photosensitizers for *in vivo* applications.
- NPs should be water soluble and should not accumulate in non-specific sites.
- NPs must not be immunogenic. They must not provoke an immune response when introduced into a tissue.
- NPs must have multifunctional properties for use in a wide range of treatments such as hypothermia and targeted radiation therapy.
- NPs must be nontoxic and stable in the biological environment.
- NPs need to interact well with cells, penetrate the cell membrane, and be well retained at the intracellular level.
- The structural integrity of the NPs should make it possible for them to selectively uptake and release biologically active molecules (photosensitizers).

PNPs can be modulated to improve their effectiveness in PDT. For example, excitation of CdSe quantum dots at 488 nm by photo irradiation generates singlet molecular oxygen ($^1O_2$) at 1270 nm. However, the poor $^1O_2$ yield (~5%) of the quantum dots suggests that it is not practical to use them due to their broad absorption band. A study demonstrated that the PDT agent can be sensitized by quantum dots through a Forster resonance energy transfer (FRET) process, or the PDT agent can interact with available molecular oxygen through a triplet energy transfer (TET) process, which results in the generation of ROS and leads to cell death. The other issues are the potential instability and cellular toxicity of Cd-based quantum dots, which limit the use of such quantum dots in PDT. Chemical structural modulation can be performed in such systems; phytochelatin peptide-based covalent conjugation of photosensitizer drug molecules with quantum dots improved the photophysical and colloidal properties of the conjugated PNPs. To continue improving the efficacy of PDT, self-lightning NPs were proposed in a recent study by Chen and Zhang. They describe an integrated approach involving radiation therapy and PDT, and it's been named as self-lightning photodynamic therapy (SLPDT). In this approach, an external light source is not required. The NPs use scintillation luminescence to activate the photosensitizer drug molecules by ionizing irradiation, which results in the generation of singlet oxygen species. However, there is no report available to show the successful direct application of these NPs in biological systems.

The other classes of NPs are biodegradable nanoparticles. They are generally made of polymers and are often enzymatically hydrolyzed in the biological environment and show controlled release of photosensitizer. The controlled release of photosensitizer depends on the molecular weight, surface charge and hydrophobicity of the polymeric nanoparticle and the pH of the environment. Among several polysaccharides and protein-based nanoparticles, alginate is used extensively for drug delivery applications. In PDT, a modified surfactant polymer formulation of alginate and anionic dioctyl sodium sulfosuccinate have shown an efficient encapsulation and sustained delivery of polar and weak bases such as methylene blue. The encapsulation and delivery increase ROS production in the breast cancer cell lines MCF7 and 4T1, resulting in more necrosis. Poly (lactic-co-glycolic acid) (PLGA) NPs were used to encapsulate mesotetra (hydroxyphenyl)porphyrin (mTHPP), mesotetraphenylporphyrin (mTPP), and mesotetraphenylporpholactol (mTPPPL) (Vargas et al., 2009). Hypericin-encapsulated PLGA-NPs have also been studied as a way to provide higher encapsulation efficiency and photoactivity in PDT (Zeisser-Labouèbe et al., 2006).

Non-biodegradable NPs have a different mode of action; they neither degrade nor control drug release in a way similar to biodegradable NPs. Due to their higher stability, they can be used repeatedly with sufficient activation. Non-biodegradable NPs have some attractive features that distinguish them from biodegradable NPs. They are not subject to microbial attack. The fine control of pore structures allows diffusion of molecular oxygen in and out of the particles, but does not allow transport of the photosensitizer agent. The shape, size and mono dispersibility of non-biodegradable NPs can be easily controlled in the preparation process, and the particles also show resistance to environmental shock because some of them persist due to their inert material properties. To explore the potential of non-biodegradable NPs in PDT, methylene blue embedded polyacrylamide NPs were used to kill rat C6 glioma tumor cells (Chouikrat et al., 2012). Most common non-degradable NPs are either silica based or metal based. Silica NPs encapsulated with mTHPC, photolon, protoporphyrin IX (PpIX), 2-devinyl-2-(1hexyloxyethyl) pyropheophorbide (HPPH), and 9,10-bis[4-(4-aminostyryl)styryl]anthracene (BDSA) have shown potential in PDT (Choi et al., 2014). Photosensitizers can be coupled onto the surface of metallic NPs, but not silica-based NPs. Due to the possibility of fabricating extremely small metallic NPs, a large number of photosensitizers can be loaded onto nanoparticles. The surface area to volume ratio is large in the small metallic NPs. AuNPs are well known for these applications due to their chemical inertness and minimal cytotoxicity. In a recent study, AuNPs were used as a carrier for the delivery of 5-aminolevulinate (5-ALA). PpIX accumulation was increased in fibrosarcoma tumor cells, resulting in higher ROS formation, which showed 50% greater cytotoxicity than the control. In one of the assembled nanocarriers consisting of polymer micelles of diacylphospholipid PEG co-filled with magnetic iron oxide nanoparticles photosensitizer, 2-(1-hexyloxyethyl)-2-devinyl pyropheophorbide-a (HPPH) showed excellent uptake and stability in HeLa cells. An external magnetic field was applied to the nanocarriers to direct *in vitro* delivery to tumor cells, suggesting magnetic migration control, enhanced imaging (iron oxide is known for magnetic resonance imaging), and phototoxicity. Despite the potential of metallic NPs in PDT, their biocompatibility, toxicity, reproducibility, aggregation, and non-specific interactions remain major challenges.

3.3 SMART NANOPARTICLES FOR PASSIVE AND ACTIVE GENE/DRUGS DELIVERY

Nanoparticles have great potential for targeted delivery of genes and drugs. The development of novel platforms for the effective transport and controlled release of drug molecules in the challenging microenvironment of diseased tissues of living systems is made possible by the use of nanostructured drug delivery systems, providing a wide range of useful nanoplatforms for promising applications in biotechnology and nanomedicine. Recent developments of smart nanocarriers are made of various organic (such as liposomes, dendrimers, hydrogels, and polymeric micelles and vesicles) and inorganic (such as gold, quantum dots, and mesoporous silica nanoparticles) materials. These nanoparticles can be incorporated into the carrier system and target the affected tissues in three different ways; passive targeting, active targeting and stimuli-responsive targeting. Nanocarrier systems combined with stimulus-responsive polymers (polymers that respond to stimuli such as redox potentials and temperature) can address some of the systemic and intracellular delivery barriers. Stimulation-responsive nanoparticles play a central role in gene and drug delivery (Figure 3.7).

Advances in materials science have created a variety of scaffolds that can respond to biological stimuli such as temperature, hypoxia, and pH, and changes in the stimulus level of lesioned tissue can cause smart nanoparticles to make drugs and biologically active molecules. With recent parallel advances in molecular cell biology, disease understanding, and controlled manipulation of material properties on the nanometer scale, nanotechnology raises great expectations for disease prevention, early and/or accurate diagnosis, and shows expectations for tailor-made treatments. The use of stimulus-responsive nanocarriers is interesting. Liposomes, block copolymer micelles, dendrimers, and polymer nanoparticles are the most commonly used molecular aggregates in such applications. The composition of each of these molecular aggregates is engineered to obtain nanoparticles with the desired stimulus response properties. The use of pH-sensitive polymers for NPs synthesis can result in significantly different pH values in the intracellular and extracellular environments.

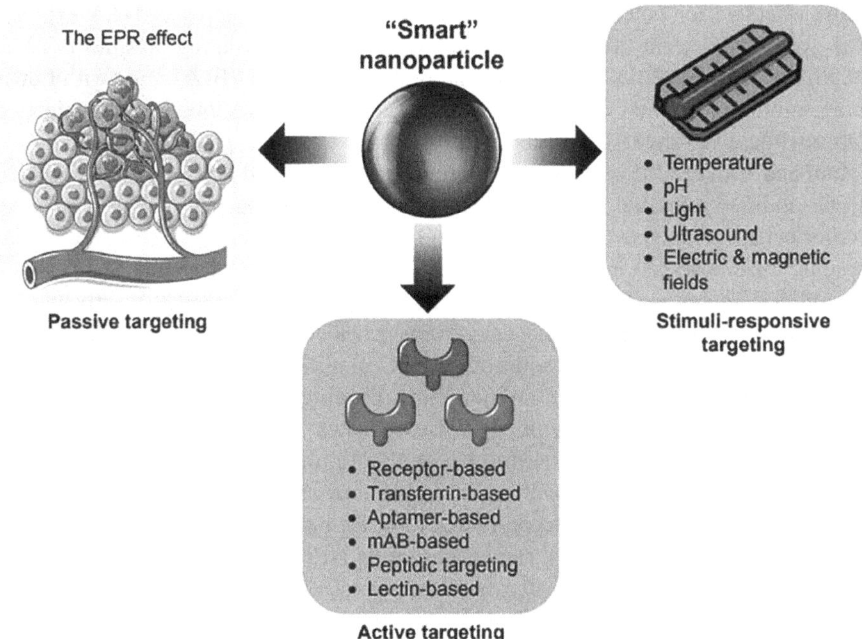

FIGURE 3.7 Multifunctional targeting employed by "smart" nanoparticles. *International Journal of Nanomedicine,* 2018, 13: 4727–4745. Originally published by and used with permission from Dove Medical Press Ltd. (Kalaydina et al., 2018.)

This is a common phenomenon in medical conditions. For example, the pH of solid tumors in the extracellular environment is more acidic (pH ~ 6.5) than blood (pH ~ 7.4) at body temperature. In addition, the pH values of intracellular endosomes and lysosomal vesicles are different from those of the cytosol. In such cases, the appropriate polymer must be selected to produce the NP. This allows nanocarriers loaded with drugs or genes to take advantage of the difference in pH to protect the cargo and successfully deliver it to the target site. Temperature is another variable available for smart delivery by setting the NP to emit the payload at a temperature higher than normal body temperature. This can encapsulate the toxic drug or gene and allow it to circulate in the bloodstream until it comes into contact with the target tissue. For example, intracellular glutathione (GSH), which is 100–1000 times higher in tumors than in the extracellular environment, can be used as a "trigger" for disulfide-bonded NPs that identify the GSH concentration gradient loaded within cell release. Normally such systems are particularly useful in providing nucleic acid-based therapies such as plasmid DNA, oligonucleotides, and small interfering RNAs. These molecules need to reach the intracellular target in a stable manner to enable effective treatment. There are several methods for successful delivery of therapeutic agents, including passive and active targeting methods. Ganta and co-workers have found an effective method of using gelatin-based NP for systemic gene delivery to solid tumors (Ganta et al., 2008). PEG-modified gelatin NP-encapsulated DNA is more effective for both *in vitro* and *in vivo* transfection of reporter DNA expressing green fluorescent protein and galactosidase by encapsulating plasmid DNA at neutral pH using gelatin NP. Systemic administration to C57/BL67 mice with Lewis lung carcinoma showed that PEG-modified gelatin NP allows long-term transfection (approximately 96 hours) both intratumorally and after intravenous administration (Kaul and Amiji, 2005). The research groups also examined GSH levels and showed PEG-modified thiolated gelatin NPs can be used to encapsulate DNA and transfect tumor cells in response to high levels of GSH. Breast cancer cell lines, and similar results, were observed *in vivo* in sympatric tumor models (Duriez and Shah, 1997). In addition, the expressed Soluble Fms-like tyrosine kinase-1 (SFLT-1) actively suppressed tumor growth and angiogenesis (Sasagawa et al., 2020). Active targeting of affected organs relies on the addition of PEG modifications to the NP to increase circulation time and achieve passive targeting. Binding of specific receptor ligands to the surface supports cell recognition at diseased sites. To target solid tumors, several surface modifications can be used to target NPs to tumor cells or to the endothelial cells of tumor blood vessels. Tumors are highly active and rapidly proliferate, so they overexpress certain surface receptors that increase nutrient uptake for nutrients like folic acid, sugars and vitamins. Hence, designing NPs with folic acid could promote their interaction with tumor cells that overexpress folate receptors, and the folate anchored NPs could then enter the cells through endocytosis before releasing their "agent" in response to the internal stimuli (Jahan et al., 2017). Tumors are highly vascularized; hence, they express specific integrin receptors like $\alpha v\beta 3$ that can interact and bind with the arginine-glycine-aspartic acid (tripeptide) RGD sequence. This can also be used as a stimulus, and RGD modification can be used to direct NPs to tumor cells or to capillary endothelial cells. Specific peptides can be identified by the phage display used to target lesion tissue in the body, especially tumors. A study by Schluesener et al. revealed a phage display of rM13 oligopeptide phage was used as a tool for peptide selection and targeting in the pathological endothelium of brain tumors in experimental rat models (Schluesener et al., 2012). This technology has been successfully used to result in adalimumab (human antibody TNF IgG1), an FDA-approved targeted treatment used to treat rheumatoid arthritis.

Aptamer technology offers additional benefits for aggressively targeting tumors under *in vivo* conditions. The development of monoclonal antibodies against antigens found only in tumor cells is another approach to targeted treatment of the disease. Regardless of which targeting approach is used; smart polymer NP is one of the useful materials for modulation and tuning to ensure efficient delivery. Smart materials that sense the environment and act accordingly to protect the drug or release the payload can be an effective treatment.

3.4 MAGNETIC NANOPARTICLES AND THEIR APPLICATIONS IN TISSUE ENGINEERING

Magnetic nanoparticles (MNPs) such as iron, cobalt, nickel and metal oxides have important advantages for remote control of cells *in vitro* and *in vivo*. MNPs enable understanding of cellular functions and signaling pathways and help develop well-defined repair biomedical procedures (Materón et al., 2021). In advance, cell surface

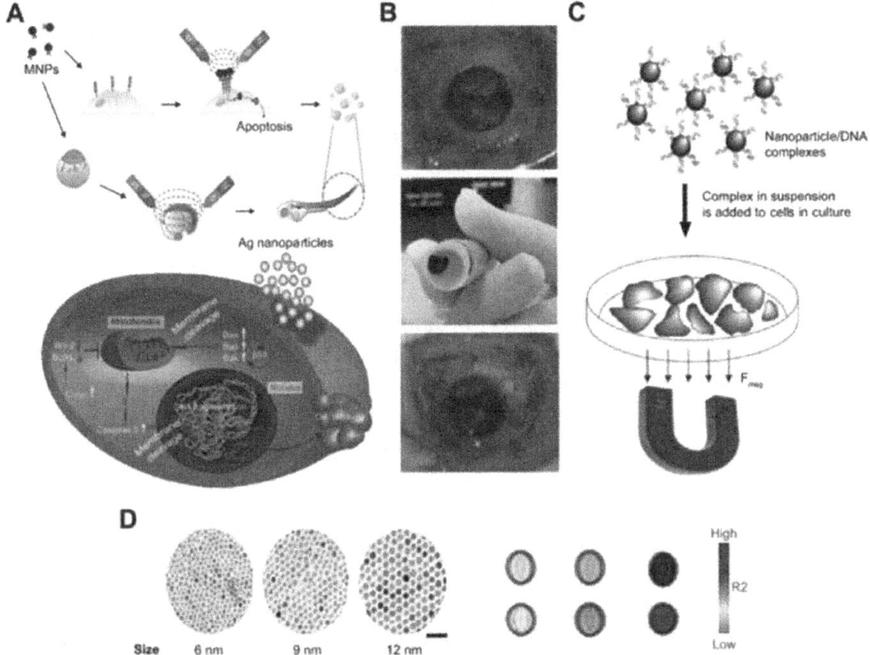

FIGURE 3.8 Utilization of magnetic particles in bioengineering applications. Notes: (A) Upper side of the figure shows MNP-induced apoptosis, whereas the bottom side shows silver nanoparticle-induced apoptosis. (B) Transplantation procedure of MSC sheets using an electromagnet with harvesting, transportation and transplantation. (C) Principles and schematic of magnetofection. (D) Size-dependent contrast agents with their T2-weighted magnetic resonance images and color maps; the scale bar is 50 nm. *International Journal of Nanomedicine,* 2018, 13: 5637–5655. Originally published by and used with permission from Dove Medical Press Ltd. (Hasan et al., 2018.)

MNP labeling or intracellular uptake serves as a probe tool for diagnostic tests and other biomedical applications. Recently, it has been discovered that superparamagnetic nanoparticles (SPM) have the potential for tissue regeneration (Figure 3.8). The single domain particle morphology of SPM (size range 15–150 nm) exhibits enhanced nanomagnetism. SPM produces an aligned magnetic moment in the presence of an electromagnetic field. It disappears completely when the electromagnetic field is removed. This behavior of SPM helps maintain colloidal stability and prevent agglomeration. However, bare MNPs exhibit a higher degree of aggregation when an external magnetic field is applied due to the magnetization of each particle. Therefore, surface functionalization is required to reduce the surface energy of MNPs and maintain their stability. Many non-polymeric organic stabilizers such as alkanesulfonic acid, hexadecylphosphonic acid, cetyltrimethylammonium bromide (CTAB), dodecylphosphonic acid, oleic acid, dihexadecylphosphonic acid, and alcanphosphonic acid have been used for surface modification of MNPs to improve the stability of MNP.

It is important that MNPs are more advantageous to mass material when they are functionalized to meet tissue or cells and are used in the subtitle range (nanoscale). Polymeric substrates such as polyethylene amine (PEI), polyvinyl alcohol (PVA), polyethylene glycol (PEG), polyvinyl acrylic acid (PAA), polyvinyl pyrrolidone (PVP), etc., also use to functionalize and stabilize MNPs. Gold and silica are also used for the stabilization of MNPs in several biological applications (Dosekova et al., 2017). MNPs is one of the clinically approved nanoparticles, which has huge applications as a therapeutic agent. Although there are some drawbacks in MNP, scientists have overcome these issues by improving the structural characteristics of MNP and combining it with tissue engineering. Tissue engineering combines biology and engineering research to repair damaged tissue by combining three major alternatives: cells, growth factors, and scaffolds. Recent advances in MNPs have mainly focused on applications in 3D tissue engineering. External cell engineering, assisted by external magnetic forces to form functional tissue, is often referred to as the "MagTE" cell membrane. This effect of magnetism has been successfully used against MNP-labeled keratinocyte cells, which are grown *in vitro* in layered epithelial grafts or integrated dermal-epidermal analogs for clinical use. In a recent study, Ito et al. claimed to be the first to report MNP-tagged keratinocytes functionalized for a multi-layered collection of undifferentiated keratinocytes (Ito et al., 2007). They also conducted an interesting study in which human beta-defensin-3 (HBD-3) was introduced into human keratinocytes and designed into a multi-layer sheet as a novel MNP-assisted antibacterial construct derived from skin tissue. Various other studies have demonstrated MNP's bone tissue engineering applications, either as a stand-alone tool or through integration with polymer substrates. Qian et al. reported the effect of magnetic NPs on bone marrow mesenchymal stem cells (Qian et al., 2018). The results indicate that the magnetic NPs are accumulated in mesenchymal stem cells with an average concentration of 20 pg per cell, which has showed no adverse effect on the proliferation and differentiation of bone marrow mesenchymal stem cells (Qian et al., 2018). The magnetic NPs promote the proliferation of bone marrow mesenchymal stem cells significantly under the action of an external magnetic field. Further, Dabrowska studied the effects of magnetic NPs and an applied magnetic field on human mesenchymal stem cells (Dabrowska et al., 2018). Its reported that after 21 days of cultivation *in vitro* with induction medium, human mesenchymal stem cells have the ability to differentiate into osteoblasts, adipocytes, and chondrocytes under the action of MNPs with the magnetic field. Although these approaches are promising and show great potential *in vitro*, they have not yet been used to develop long-lived, angiogenic, natural tissue constructs for clinical use (Fan et al., 2020).

REFERENCES

Abu-Salah, Khalid M, Mohammed M Zourob, Fouzi Mouffouk, Salman A Alrokayan, Manal A Alaamery, and Anees A Ansari. 2015. "DNA-based nanobiosensors as an emerging platform for detection of disease." *Sensors* no. 15 (6):14539–14568.

Auyeung, Evelyn, Ting ING Li, Andrew J Senesi, Abrin L Schmucker, Bridget C Pals, Monica Olvera de La Cruz, and Chad A Mirkin. 2014. "DNA-mediated nanoparticle crystallization into Wulff polyhedra." *Nature* no. 505 (7481):73–77.

Bohunicky, Brian, and Shaker A Mousa. 2010. "Biosensors: The new wave in cancer diagnosis." *Nanotechnology, Science and Applications* no. 4:1–10.

Choi, Bong Gill, HoSeok Park, Tae Jung Park, Min Ho Yang, Joon Sung Kim, Sung-Yeon Jang, Nam Su Heo, Sang Yup Lee, Jing Kong, and Won Hi Hong. 2010. "Solution chemistry of self-assembled graphene nanohybrids for high-performance flexible biosensors." *ACS Nano* no. 4 (5):2910–2918.

Choi, Soonmo, Anuj Tripathi, and Deepti Singh. 2014. "Smart nanomaterials for biomedics." *Journal of Biomedical Nanotechnology* no. 10 (10):3162–3188.

Chouikrat, Rima, Aymeric Seve, Régis Vanderesse, Hamanou Benachour, Muriel Barberi-Heyob, Sébastien Richeter, Laurence Raehm, J-O Durand, Marc Verelst, and Céline Frochot. 2012. "Non polymeric nanoparticles for photodynamic therapy applications: Recent developments." *Current Medicinal Chemistry* no. 19 (6):781–792.

Dabrowska, Sylwia, Andrea Del Fattore, Elzbieta Karnas, Malgorzata Frontczak-Baniewicz, Hanna Kozlowska, Maurizio Muraca, Miroslaw Janowski, and Barbara Lukomska. 2018. "Imaging of extracellular vesicles derived from human bone marrow mesenchymal stem cells using fluorescent and magnetic labels." *International Journal of Nanomedicine* no. 13:1653.

Dhara, Keerthy, John Stanley, T Ramachandran, Bipin G Nair, and Satheesh Babu TG. 2014. "Pt-CuO nanoparticles decorated reduced graphene oxide for the fabrication of highly sensitive non-enzymatic disposable glucose sensor." *Sensors and Actuators B: Chemical* no. 195:197–205.

Ding, Lin, Chen Hao, Yadong Xue, and Huangxian Ju. 2007. "A bio-inspired support of gold nanoparticles– chitosan nanocomposites gel for immobilization and electrochemical study of K562 Leukemia cells." *Biomacromolecules* no. 8 (4):1341–1346.

Dosekova, Erika, Jaroslav Filip, Tomas Bertok, Peter Both, Peter Kasak, and Jan Tkac. 2017. "Nanotechnology in glycomics: Applications in diagnostics, therapy, imaging, and separation processes." *Medicinal Research Reviews* no. 37 (3):514–626.

Duriez, Patrick, and Girish M Shah. 1997. "Cleavage of poly (ADP-ribose) polymerase: A sensitive parameter to study cell death." *Biochemistry and Cell Biology* no. 75 (4):337–349.

Elghanian, Robert, James J Storhoff, Robert C Mucic, Robert L Letsinger, and Chad A Mirkin. 1997. "Selective colorimetric detection of polynucleotides based on the distance-dependent optical properties of gold nanoparticles." *Science* no. 277 (5329):1078–1081.

Fan, Daoyang, Qi Wang, Tengjiao Zhu, Hufei Wang, Bingchuan Liu, Yifan Wang, Zhongjun Liu, Xunyong Liu, Dongwei Fan, and Xing Wang. 2020. "Recent advances of magnetic nanomaterials in bone tissue repair." *Frontiers in Chemistry* no. 8:745.

Fogel, Ronen, Janice Limson, and Ashwin A Seshia. 2016. "Acoustic biosensors." *Essays in Biochemistry* no. 60 (1):101–110.

Ganta, Srinivas, Harikrishna Devalapally, Aliasgar Shahiwala, and Mansoor Amiji. 2008. "A review of stimuli-responsive nanocarriers for drug and gene delivery." *Journal of Controlled Release* no. 126 (3):187–204.

Gao, Zhuangqiang, Mingdi Xu, Li Hou, Guonan Chen, and Dianping Tang. 2013. "Magnetic bead-based reverse colorimetric immunoassay strategy for sensing biomolecules." *Analytical Chemistry* no. 85 (14):6945–6952.

Gray, Mark, James Meehan, Carol Ward, Simon P Langdon, Ian H Kunkler, Alan Murray, and David Argyle. 2018. "Implantable biosensors and their contribution to the future of precision medicine." *The Veterinary Journal* no. 239:21–29.

Grieshaber, Dorothee, Robert MacKenzie, Janos Vörös, and Erik Reimhult. 2008. "Electrochemical biosensors-sensor principles and architectures." *Sensors* no. 8 (3):1400–1458.

Gui, Qingyuan, Tom Lawson, Suyan Shan, Lu Yan, and Yong Liu. 2017. "The application of whole cell-based biosensors for use in environmental analysis and in medical diagnostics." *Sensors* no. 17 (7):1623.

Hasan, Anwarul, Mahboob Morshed, Adnan Memic, Shabir Hassan, Thomas J Webster, and Hany El-Sayed Marei. 2018. "Nanoparticles in tissue engineering: Applications, challenges and prospects." *International Journal of Nanomedicine* no. 13:5637.

Ho, Mei-Lin, Jing-Chang Wang, Ting-Yi Wang, Chun-Yen Lin, Jian Fan Zhu, Yi-An Chen, and Tsai-Chen Chen. 2014. "The construction of glucose biosensor based on crystalline iridium (III)-containing coordination polymers with fiber-optic detection." *Sensors and Actuators B: Chemical* no. 190:479–485.

Hua, Yu, Jiaming Ma, Dachao Li, and Ridong Wang. 2022. "DNA-based biosensors for the biochemical analysis: A review." *Biosensors* no. 12 (3):183.

Huang, Xiaohua, and Mostafa A El-Sayed. 2010. "Gold nanoparticles: Optical properties and implementations in cancer diagnosis and photothermal therapy." *Journal of Advanced Research* no. 1 (1):13–28.

Ishikawa, Fumiaki N, Hsiao-Kang Chang, Marco Curreli, Hsiang-I Liao, C Anders Olson, Po-Chiang Chen, Rui Zhang, Richard W Roberts, Ren Sun, and Richard J Cote. 2009. "Label-free, electrical detection of the SARS virus N-protein with nanowire biosensors utilizing antibody mimics as capture probes." *ACS Nano* no. 3 (5):1219–1224.

Ito, Akira, Hideaki Jitsunobu, Yoshinori Kawabe, and Masamichi Kamihira. 2007. "Construction of heterotypic cell sheets by magnetic force-based 3-D coculture of HepG2 and NIH3T3 cells." *Journal of Bioscience and Bioengineering* no. 104 (5):371–378.

Jahan, Sheikh Tasnim, Sams Sadat, Matthew Walliser, and Azita Haddadi. 2017. "Targeted therapeutic nanoparticles: An immense promise to fight against cancer." *Journal of Drug Delivery* no. 2017:9090325.

Jeong, Bongjin, Rashida Akter, Oc Hee Han, Choong Kyun Rhee, and Md Aminur Rahman. 2013. "Increased electrocatalyzed performance through dendrimer-encapsulated gold nanoparticles and carbon nanotube-assisted multiple bienzymatic labels: Highly sensitive electrochemical immunosensor for protein detection." *Analytical Chemistry* no. 85 (3):1784–1791.

Kalaydina, Regina-Veronicka, Komal Bajwa, Bessi Qorri, Alexandria Decarlo, and Myron R Szewczuk. 2018. "Recent advances in "smart" delivery systems for extended drug release in cancer therapy." *International Journal of Nanomedicine* no. 13:4727.

Karunakaran, Chandran, Raju Rajkumar, and Kalpana Bhargava. 2015. "Introduction to biosensors." In Chandran Karunakaran, Kalpana Bhargava, and Robson Benjamin (Eds.), *Biosensors and bioelectronics*, pp. 1–68. Elsevier, Amsterdam.

Kaul, Goldie, and Mansoor Amiji. 2005. "Tumor-targeted gene delivery using poly (ethylene glycol)-modified gelatin nanoparticles: In vitro and in vivo studies." *Pharmaceutical Research* no. 22 (6):951–961.

Khalil, Ibrahim, Nurhidayatullaili Muhd Julkapli, Wageeh A Yehye, Wan Jefrey Basirun, and Suresh K Bhargava. 2016. "Graphene–gold nanoparticles hybrid—synthesis, functionalization, and application in a electrochemical and surface-enhanced raman scattering biosensor." *Materials* no. 9 (6):406.

Kim, Hyung Shik, and Dong Yun Lee. 2018. "Near-infrared-responsive cancer photothermal and photodynamic therapy using gold nanoparticles." *Polymers* no. 10 (9):961.

Kulkarni, Madhusudan B, Narasimha H Ayachit, and Tejraj M Aminabhavi. 2022. "Biosensors and microfluidic biosensors: From fabrication to application." *Biosensors* no. 12 (7):543.

Law, Wing-Cheung, Ken-Tye Yong, Alexander Baev, and Paras N Prasad. 2011. "Sensitivity improved surface plasmon resonance biosensor for cancer biomarker detection based on plasmonic enhancement." *ACS Nano* no. 5 (6):4858–4864.

Li, Songjun, Yi Ge, and He Li. 2012. *Smart Nanomaterials for Sensor Application*. Bentham Science Publishers, Bussum.

Liu, Guodong, Xun Mao, Joseph A Phillips, Hui Xu, Weihong Tan, and Lingwen Zeng. 2009. "Aptamer– nanoparticle strip biosensor for sensitive detection of cancer cells." *Analytical Chemistry* no. 81 (24):10013–10018.

Loretan, Morgane, Ivana Domljanovic, Mathias Lakatos, Curzio Rüegg, and Guillermo P Acuna. 2020. "DNA origami as emerging technology for the engineering of fluorescent and plasmonic-based biosensors." *Materials* no. 13 (9):2185.

Luo, Daibing, Liangzhuan Wu, and Jinfang Zhi. 2009. "Fabrication of boron-doped diamond nanorod forest electrodes and their application in nonenzymatic amperometric glucose biosensing." *ACS Nano* no. 3 (8):2121–2128.

Malhotra, Bansi Dhar, and Md Azahar Ali. 2018. "Nanomaterials in biosensors: Fundamentals and applications." Nanomaterials for *B*iosensors, 1–73.

Materón, Elsa M, Celina M Miyazaki, Olivia Carr, Nirav Joshi, Paulo HS Picciani, Cleocir J Dalmaschio, Frank Davis, and Flavio M Shimizu. 2021. "Magnetic nanoparticles in biomedical applications: A review." *Applied Surface Science Advances* no. 6:100163.

Ning, Yan-Na, Bao-Lin Xiao, Nan-Nan Niu, Ali Akbar Moosavi-Movahedi, and Jun Hong. 2019. "Glucose oxidase immobilized on a functional polymer modified glassy carbon electrode and its molecule recognition of glucose." *Polymers* no. 11 (1):115.

North, John R. 1985. "Immunosensors: Antibody-based biosensors." *Trends in Biotechnology* no. 3 (7):180–186.

Pandey, Pratibha, Monika Datta, and BD Malhotra. 2008. "Prospects of nanomaterials in biosensors." *Analytical Letters* no. 41 (2):159–209.

Qian, Wenbo, Min Qian, Yi Wang, Jianfei Huang, Jian Chen, Lanchun Ni, Qingfeng Huang, Qianqian Liu, Peipei Gong, and Shiqiang Hou. 2018. "Combination glioma therapy mediated by a dual-targeted delivery system constructed using OMCN–PEG–Pep22/DOX." *Small* no. 14 (42):1801905.

Rocchitta, Gaia, Angela Spanu, Sergio Babudieri, Gavinella Latte, Giordano Madeddu, Grazia Galleri, Susanna Nuvoli, Paola Bagella, Maria Ilaria Demartis, and Vito Fiore. 2016. "Enzyme biosensors for biomedical applications: Strategies for safeguarding analytical performances in biological fluids." *Sensors* no. 16 (6):780.

Sagle, Laura B, Laura K Ruvuna, Julia M Bingham, Chunming Liu, Paul S Cremer, and Richard P Van Duyne. 2012. "Single plasmonic nanoparticle tracking studies of solid supported bilayers with ganglioside lipids." *Journal of the American Chemical Society* no. 134 (38):15832–15839.

Sasagawa, Tadashi, Atsushi Jinno-Oue, Takeshi Nagamatsu, Kazuki Morita, Tetsushi Tsuruga, Mayuyo Mori-Uchino, Tomoyuki Fujii, and Masabumi Shibuya. 2020. "Production of an anti-angiogenic factor sFLT1 is suppressed via promoter hypermethylation of FLT1 gene in choriocarcinoma cells." *BMC Cancer* no. 20 (1):1–13.

Schluesener, Hermann Josef, Yanhua Su, Azadeh Ebrahimi, and Davoud Pouladsaz. 2012. "Antimicrobial peptides in the brain: Neuropeptides and amyloid." *Frontiers in Bioscience-Scholar* no. 4:1375–1380.

Scholten, Kee, and Ellis Meng. 2018. "A review of implantable biosensors for closed-loop glucose control and other drug delivery applications." *International Journal of Pharmaceutics* no. 544 (2):319–334.

Sin, Mandy LY, Kathleen E Mach, Pak Kin Wong, and Joseph C Liao. 2014. "Advances and challenges in biosensor-based diagnosis of infectious diseases." *Expert Review of Molecular Diagnostics* no. 14 (2):225–244.

Thavanathan, Jeevan, Nay Ming Huang, and Kwai Lin Thong. 2014. "Colorimetric detection of DNA hybridization based on a dual platform of gold nanoparticles and graphene oxide." *Biosensors and Bioelectronics* no. 55:91–98.

Vargas, Angelica, Norbert Lange, Tudor Arvinte, Radovan Cerny, Robert Gurny, and Florence Delie. 2009. "Toward the understanding of the photodynamic activity of m-THPP encapsulated in PLGA nanoparticles: Correlation between nanoparticle properties and in vivo activity." *Journal of Drug Targeting* no. 17 (8):599–609.

Yin, Guang, Ling Xing, Xiu-Ju Ma, and Jun Wan. 2014. "Non-enzymatic hydrogen peroxide sensor based on a nanoporous gold electrode modified with platinum nanoparticles." *Chemical Papers* no. 68 (4):435–441.

Zeisser-Labouèbe, Magali, Norbert Lange, Robert Gurny, and Florence Delie. 2006. "Hypericin-loaded nanoparticles for the photodynamic treatment of ovarian cancer." *International Journal of Pharmaceutics* no. 326 (1–2):174–181.

Zhang, Guangxun, Xiao Xiao, Bing Li, Peng Gu, Huaiguo Xue, and Huan Pang. 2017. "Transition metal oxides with one-dimensional/one-dimensional-analogue nanostructures for advanced supercapacitors." *Journal of Materials Chemistry A* no. 5 (18):8155–8186.

Zhou, Guobao, Meihua Lin, Ping Song, Xiaoqing Chen, Jie Chao, Lianhui Wang, Qing Huang, Wei Huang, Chunhai Fan, and Xiaolei Zuo. 2014. "Multivalent capture and detection of cancer cells with DNA nanostructured biosensors and multibranched hybridization chain reaction amplification." *Analytical Chemistry* no. 86 (15):7843–7848.

4 SNM for Pharmaceutical Applications and Analysis

4.1 PHYSICAL RESPONSIVE NANOMATERIALS

Physical stimuli-responsive SNM are intelligent materials capable of controlling drug release in response to the various physical stimuli such as magnetic field, temperature, ultrasound, light, and electric field. Over the last few decades, many strategies have been developed to program them to have multiple functionalities, a low degree of variability, and a high precision to address the unmet need for on-demand and targeted drug delivery. These strategies are classified into three categories: chemistry (including basic/core chemistry and the chemistry of surface targeting ligands such as antibodies, peptides, and aptamers, among others), architecture of nanomaterials, and physical stimuli parameters such as type, intensity, and duration, among others. All of these strategies can be used to control nanomaterials' interactions with drugs, and drug loading and drug-release efficiency. These strategies can also be used to manipulate the uptake of nanotherapeutics by cells and tissues, as well as their permeability across biological barriers, which indicate the targeting effect. However, several major obstacles must be overcome before these physical stimuli-responsive nanomaterials can be successfully translated into the clinical practices. The first challenge is to keep these nanomaterials from accumulating and/or being taken up by non-target tissues. Off-target accumulation/uptake is primarily caused by non-specific protein adsorption on nanomaterial surfaces (forming a protein corona) in the biological milieu (Akhter et al., 2021).

Protein adsorption on nanomaterials frequently results in protein denaturation, which initiates a signal cascade that results in nanomaterial aggregation and/or phagocytosis via activated macrophages (Shemetov et al., 2012). Since the protein adsorption is non-specific, it can also happen to nanomaterials targeting moieties. Consequently, the protein adsorption causes more nanomaterials to accumulate in organs such as kidney, liver, spleen, etc., rather than the required target sites. The second challenge that these stimuli-responsive nanomaterials share with conventional nanotherapeutics is the inefficiency with which the nanotherapeutics are cleared from the body once they have completed their mission. Since most nanotherapeutics are larger than the renal threshold, they cannot be removed from the body through the kidneys; they tend to accumulate in the body if they are not biodegradable.

Even for some biodegradable nanomaterials, degraded fragments may be trapped in lysosomal compartments, causing toxicity/side effects. The third challenge is that the most targeting moieties conjugated on nanomaterials are not specific to the target sites because receptors for the targeting moieties are expressed not only at the target sites but also in other organs. For example, the folate receptor is overexpressed in many cancers, but it is also expressed at a moderate to high level in

DOI: 10.1201/9781003366270-4

normal organs such as the small intestine, placenta, and kidneys (Sahle et al., 2018). Furthermore, the overexpressed folate receptor is inhomogeneously distributed on malignant cells, resulting in non-uniform nanotherapeutic accumulation in the target tissue. Furthermore, some targeting moieties, such as antibodies and peptides, may lose activity during nanomaterial conjugation and fail to induce the desired tissue targeting effect. Targeting ligands on the surface of nanocarriers may also change the surface properties of nanomaterials such as their charge and hydrophobicity, resulting in increased opsonization, aggregation, and clearance of the nanomaterials by the mononuclear-phagocyte system. The fourth challenge is that some physical stimuli may not be fully tolerated by the body, making their use and regulation\more expensive. Ultraviolet (UV) light, for example, cannot penetrate deeper than 10mm into tissues due to absorption by endogenous chromophores such as lipids, oxy- and deoxy-hemoglobin, and water; and prolonged UV irradiation can be cytotoxic. As a result, UV-responsive nanotherapeutics should be limited to the eye, skin, and other mucosal surfaces, be doped with up conversion luminescent materials, or be used in conjunction with near infrared (NIR). The cavitation caused by an ultrasound stimulus may increase cancer cell vessel permeability, resulting in metastatic spread. Electrical stimulation has low tissue penetration and may cause tissue damage, limiting the clinical application of electro-responsive nanoparticles despite their flexibility and low cost. Magnetic field stimulation is expensive due to its complexity and the need for specialized equipment for adequate focusing and deep penetration into the disease area with sufficient strength.

Thermoresponsive materials require more time for transition into a phase that results in burst drug release and precise temperature control at the target site without causing tissue damage. Due to these difficulties, only a small number of physical stimuli-responsive nanotherapeutics have progressed to clinical trials. For example, NIR-absorbing nanomaterials are under test in different clinical trials, promising applicability for the treatment of several types of cancer and metastases in the near future. For instance, the NIR-absorbing nanosystem AuroLase obtained the investigational device exemption, and is currently studied in a human clinical trial for the treatment of refractory/recurrent head and neck cancer (NIH Clinical Trial NCT000848042), and of primary and/or metastatic lung tumors (NIH Clinical Trial NCT01679470). A more recent clinical trial from the same AuroLase provider relies on a magnetic resonance imaging ultrasound (MRI/US) fusion biopsy in conjunction with PTA for the treatment of prostate cancer (NIH Clinical Trial NCT02680535) (Genchi et al., 2017).

As a result, continuous design improvements, more in vivo toxicology and efficacy evaluations, and robust stability and production scale-up studies on these nanomaterials are expected in the future in order for physical stimuli-responsive nanotherapeutics to be developed into intelligent drug-delivery systems to treat human diseases.

4.2 TEMPERATURE-RESPONSIVE NANOMATERIALS

Temperature- or thermoresponsive systems are among the most extensively investigated exogenous stimuli-responsive systems for treatment and diagnostic applications (Sánchez-Moreno et al., 2018). Thermosensitive nanomaterials are a type of smart material that is particularly important in biomedical applications. In most cases, the thermoresponsive nature of the nanomaterial is driven by its polymeric nature, although liposomes have also been reported to be thermosensitive. Hydrogels are the most commonly used thermoresponsive polymers. The thermoresponsive property of the polymeric constituent is based on either critical solution behavior or shape memory (Sánchez-Moreno et al., 2018).

When the polymer chain dehydrates in response to an external temperature change above the critical solution temperature, the nanocarriers can release their drug payload. Diseased/tumor tissues generally have a higher temperature (40 °C–42 °C) than normal tissues (37 °C) (Liu et al., 2016). As a result, thermoresponsive drug carriers retain their drug payload at normal temperatures and release it when exposed to the diseased/tumor tissue's elevated temperatures.

There are two approaches to thermoresponsive drug release. One approach is to use drug carriers that are designed to release drugs in bursts when exposed to high temperatures. Polycaprolactone(*N*-isopropyl acrylamide) (PNIPAm) conjugated carbon nanomaterials, for example, have been prepared with drug molecules to treat tumors (Sánchez-Moreno et al., 2018). These drug-loaded nanoparticles were thermoresponsive and released more drugs at higher temperatures (40 °C). Furthermore, after treatment with the drug-loaded nanoparticles, cell viability (20%) at 40 °C was significantly lower than that at 37 °C (40%).When heated above its LCST, the popular temperature-responsive polymer PNIPAm undergoes a reversible phase transition, which has been extensively used as a scaffold to control the distance and coupling of plasmonic nanoparticles. Several approaches have been taken, including coating nanoparticles with PNIPAm, synthesis of the nanoparticles in situ within the PNIPAm matrix, and tethering the nanoparticles to the surface of PNIPAm spheres. This control of plasmonic nanostructure structural parameters is critical for their applications, particularly in spectroscopy and biosensing. The most commonly studied combination is PNIPAm with gold nanosystems (Figure 4.1).

The second approach uses drug carriers that are designed to release drugs in a burst when exposed to higher temperatures caused by an external stimulus. The external stimulus causes a sensitive drug carrier agent to generate heat, which causes temperature-sensitive materials in the drug carrier to change, resulting in burst drug release at the target site (Khoee and Karimi, 2018). Antitumor drug molecules were loaded into an amphiphilic di-block copolymer prepared with temperature-responsive materials such as polypyrrole (PPy). Because of the photothermal effect, PPy produced heat upon NIR absorption, and the increase in temperature promoted drug release from the micelles. The swelling of the thermosensitive polymer caused the release via hydrophobic to hydrophilic conversion (Yang et al., 2018b).

Because they can respond to temperature changes, thermoresponsive polymers are critical components of these systems. Thermoresponsive materials have two distinct properties: the lower CST (critical solution temperature) (LCST) and the upper CST (critical solution temperature) (UCST). Swelling occurs when the temperature falls below the LCST due to increased hydrophilicity, and swelling occurs when the temperature rises above the UCST due to increased hydrophilicity. Changes in hydrophilicity control the swelling behavior of drug carriers and enable drug release to be fine-tuned. For example, PNIPAM is a common building block for thermoresponsive

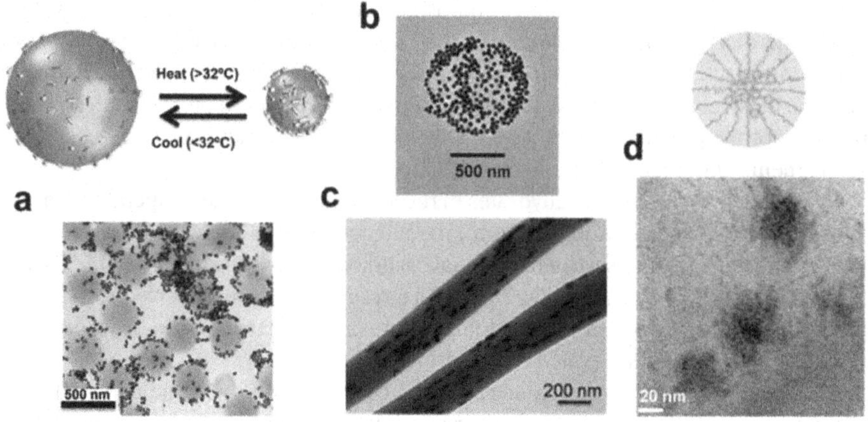

FIGURE 4.1 Examples of Au-PNIPAm-based thermosensitive nanosystems (a) Schematic representation and TEM image of Au nanorods (NRs)–PNIPAm microgel hybrids and the corresponding thermoresponsive behavior (b) TEM image of PNIPAm–Au composite (c) TEM image of AuNRs/PNIPAm electrospun fibers. (d) Schematic representation and TEM image of Au-thermosensitive polymer (PEG-PNIPAm) nanohybrids. (Sánchez-Moreno et al., 2018.)

carriers (Kim and Matsunaga, 2017). Its solubility in water varies with temperature variation from its LCST. PNIPAM coils transform into water-insoluble globules above their LCST; thus, drug release can be controlled due to the dominance of hydrophobic interactions. Hydrophilic PNIPAM was used below its LCST for higher drug loading, while hydrophobic PNIPAM was used above its LCST for cell attachment and sustained release. Furthermore, its LCST can be altered by varying the ratio of hydrophilic to hydrophobic components (Park et al., 2016). Moreover, Munaweera et al. investigated the combination of ciprofloxacin-loaded temperature-sensitive liposomes and alternating magnetic fields (AMFs). Upon combining AMFs with temperature-sensitive liposomal ciprofloxacin, a 3 log reduction in colony forming units (CFU) of *Pseudomonas aeruginosa* in biofilm was observed in metal washers (Munaweera et al., 2018).

4.3 ELECTRICAL AND ELECTROCHEMICAL STIMULI-RESPONSIVE NANOMATERIALS

Electric fields are one of the many physical triggers that have helped to improve the efficacy of drug therapy. Exogenous high-intensity electric fields can directly influence the permeability of cellular membranes, be used as a stimulus for drug delivery, or be widely used as therapeutic tools for wound healing and tissue integrity restoration. Promising electrochemical stimuli-responsive nanomaterials for controlled drug release or bioactive scaffolds include conductive polymers and hydrogel-based materials.

When a potential is applied, the conducting polymer is reduced which is the mechanism underlying this approach. Because of electrostatic repulsion, the drug incorporated into the polymer is discharged into the surrounding environment. The rate of drug release is determined by the morphology (including density) of the polymer, its electrochemical properties, or the properties of the solution surrounding the drug-delivery system (e.g., pH, temperature, and the addition of the dopants). Neumann et al. recently published a strategy for electro-responsive drug delivery (Neumann et al., 2018). By synthesizing drug-loaded nanofilms, they have employed local pH changes caused by an electrochemical reaction, and the pH change caused by the electrical signal recovered quickly after the stimulus was removed due to a buffering action, which prevented an "off" state release.

Electro-responsive hydrogels are water-swollen macromolecules (networks of hydrophilic polyelectrolyte chains with a high number of ionizable groups along the polymer backbone) that can swell or shrink in response to electrical stimuli. The application of electro-responsive hydrogels in therapeutic approaches is still in its early stages, but it holds great promise in electro-stimulated drug delivery. When an electric field is applied to polysaccharide hydrogels, changes in polarity, ionic strength, and pH alter the net osmotic pressure within polymers (cause electro-osmosis), resulting in polymer bending, swelling, shrinking, or erosion and subsequent release of the active ingredient (Kolosnjaj-Tabi et al., 2019).

4.4 LIGHT-RESPONSIVE NANOMATERIALS

Specific light wavelengths (ultraviolet, visible, and near-infrared light) can alter the stability or structural degradation of responsive nanomaterials in light-responsive drug-delivery systems, allowing drugs to be released at precise locations.

Visible and UV light are not suitable for *in vivo* therapeutic applications due to their low penetration capability. A NIR-responsive system, on the other hand, is a promising technique that uses light to control drug release because it has better penetration (13 nm) and causes less tissue damage (Xiang et al., 2018).

In NIR-responsive systems, three different drug-release mechanisms are used: the photothermal effect, two-photon absorption (TPA), and the upconverting nanoparticles (UCNPs) mechanism.

The photothermal effect is caused by a photothermal agent contained in the nanocarrier converting light to heat. This heat stimulates the heat-sensitive material, disrupting the nanostructure, or creating a phase transition that leads to rapid drug release at the tumor site. Li et al. recently created a number of nanostructured lipid carriers encapsulated by liposomes that contained the hydrophilic drug AMD3100 and the hydrophobic NIR photothermal agent IR780 (Li et al., 2017). NIR light stimulated IR780, causing heat to be produced, which destabilized the liposomal membrane and caused drug release. In addition to chemotherapy, IR780 induced cytotoxic hyperthermia as a synergistic effect.

TPA is based on the excitation of two absorbed photons of different or identical frequencies. (Yang et al., 2016). Guardado-Alvarez et al., for example, created mesoporous silica nanoparticles (MSNPs) with a disulfide-linked-cyclodextrin cap (Guardado-Alvarez et al., 2014). The TPA-based transducer supplied an electron for the disulfide linker reduction. This resulted in the removal of the -cyclodextrin cap, resulting in drug release. TPA is a promising strategy for controlled drug delivery because NIR lasers have high spatial and temporal resolutions, deep tissue penetration, and low scattering losses. To treat a small infection area, this technique requires a focal pulsed laser with a high energy density. As a result, this method is not suitable for in vivo experiments.

The UCNPs technique can convert NIR light to UV light, allowing high-energy light-sensitive materials to be activated (Gwon et al., 2018). Xiang et al., for example, synthesized and coated UCNPs with an amphiphilic di-block copolymer containing a UV-sensitive hydrophobic layer made of poly(4,5-dimethoxy-2-nitrobenzyl methacrylate) and an outer hydrophilic layer made of poly(4,5-dimethoxy-2-nitrobenzyl methacrylate) (methoxy polyethylene glycol monomethacrylate) (Xiang et al., 2018). When exposed to NIR irradiation (908 nm), a copolymer absorbed the UV light and caused an imbalance in the hydrophilic-hydrophobic equilibrium, causing the micelle structure to change and the encapsulated drug to be released. poly(4,5-dimethoxy-2-nitrobenzyl methacrylate) (PNB) absorbed UV light and converted a hydrophobic polymer block into a hydrophilic block, resulting in the dissolution of a di-block polymer and the release of drug molecules (Li et al., 2018).

Huang et al. discovered that ultrathin InSe nanosheets were effective NIR-II drug-release agents (Huang et al., 2019). PEGylated InSe nanosheets demonstrated excellent biocompatibility and physiological stability, as well as a 39.5% NIR-II photothermal conversion efficiency and a 93.6% (w/w) DOX-loading capacity. The on/off switching stimulus accelerated the release rate of DOX under NIR-II irradiation. This suggests that NIR-II-induced local hyperthermia can release DOX from InSe-DOX in a controlled and time-dependent manner. DOX release can be increased by irradiation in acidic conditions. Thus, the InSe nanoplatforms' pH responsiveness, NIR-II sensitivity, and switchable release characteristics make it useful as a targeted chemotherapy delivery method.

4.5 MAGNETIC-RESPONSIVE NANOMATERIALS

Magnetic systems are frequently used for body imaging (e.g., MRI) and drug-release control via external stimuli (Yang et al., 2018a). They are widely applicable due to their biocompatibility, biodegradability, ease of synthesis as a co-precipitate or micro-emulsion, and ease of modification and functionalization for specific applications. Magnetic nanoparticles (MNPs) have a small specific surface area and can easily reach desired locations. As a result, MNPs appear to be a promising drug-delivery system. Tumor cells are more sensitive to temperature increases than healthy cells, MNPs can be utilized to warm up the cancerous cells and kill those affected cells. The MNPs will damage the cancer cells by entering the tumors and then releasing energy as heat when subjected to an AMF. Most commonly used MNPs in clinical applications are ferromagnetic (such as cobalt, nickel, and iron), paramagnetic (such as magnesium, lithium, tantalum, and gadolinium), diamagnetic (such as silver, copper, gold, and the majority of known elements), antiferromagnetic (such as CoO, MnO, $CuCl_2$, and NiO), and ferrimagnetic (such as maghemite γ-Fe_2O_3 and magnetite Fe_3O_4) (Alromi et al., 2021).

The ability of magnetic stimuli-responsive drug-release systems to generate heat via an alternate magnetic frequency is central to their basic mechanism. Under an alternating current magnetic field, MNPs can act as transducers, converting hysteresis loss and Néel relaxation to heat. Two magnetic-responsive system-based treatment mechanisms have been reported. Magnetic field-induced hyperthermia is one mechanism, and magnetic field-guided drug targeting is another (Thirunavukkarasu et al., 2018).

Due to their magnetic responses, hyperthermia-based magnetic systems have been investigated for drug-delivery and tumor inhibition applications. Thirunavukkarasu et al. created theranostic superparamagnetic iron oxide (Fe_3O_4) nanoparticles (SPIONs). They placed SPIONs and DOX in a poly(lactic-co-glycolic acid) (PLGA) matrix, which reacts to the heat produced by the SPIONs when exposed to a magnetic field, releasing DOX (Lee et al., 2018). Temperature sensitivity of the PLGA matrix was demonstrated *in vitro*, with 39% and 57% drug release observed at 37 °C and 45°C, respectively. Under the external application of external AMF, iron oxide nanoparticles changed the temperature of the solution, causing the PLGA polymer matrix to transition and the drug molecules to be released. Furthermore, Munaweera and co-workers have synthesized platinum (Pt)-based chemotherapeutic (cisplatin, carboplatin, or oxaliplatin) and holmium-165 (Ho), which can be neutron-activated to produce the holmium-166 radionuclide (HoIG-Pt) for selective delivery to tumors using an external magnet (Munaweera et al., 2015).

Wang et al. created an implantable magnetic chitosan hydrogel loaded with hydrophobic (rifampicin) and hydrophilic (adriamycin) drugs in another study. The implantable magnetic chitosan hydrogel responded to an external low frequency alternating magnetic field (AMF) by pulsatile release of drugs without causing magnetic hyperthermia. This system was created to actively regulate drug release and prevent postoperative infections.

4.6 CHEMICAL-RESPONSIVE NANOMATERIALS

Polymers, phosphodiesters, and inorganic materials have been widely used to create enzyme-responsive drug-delivery systems. For example, the ester bonds or peptide structure of stimuli-responsive nanocarriers can be degraded by enzymes specific to tumor inflammation. The drug payload of these nanocarriers is released at specific sites as a result of this degradation. Biorecognition, process efficiency, sensitivity, selectivity, and catalytic efficacy are all advantages of enzyme-responsive systems. The drug payload is released in these systems via the degradation of enzyme-assisted polymeric moieties. Proteases (or peptidases) and phospholipases are two types of enzymes that are commonly used as triggers in enzyme-responsive drug delivery. Proteases are particularly useful in the development of these drug-delivery systems because they are frequently overexpressed during infection, cancer, and inflammation. Trypsin, one of the essential digestive proteinases, regulates exocrine pancreatic secretion, which stimulates several other digestive enzymes. Because of its upregulation in the tumor microenvironment, phospholipase A2 (PLA2) is gaining popularity as a therapeutic target. For example, Mahdi Ghavami et al. developed the phospholipase sensitive liposome (PSL) as a drug-delivery system in which liposome degradation is caused by phospholipase A2 secreted by tumor cells (sPLA2). sPLA2 is in-charge of causing PSL to release active PNA (peptide nucleic acid) as an antisense agent. The results showed that more than 80% of the phospholipase was used to activate PNA release. As a result, phospholipase could be used as a trigger agent to release drugs from nano-delivery platforms.

An enzyme-responsive nanomaterial was created by modifying an *N*-(2-hydroxypropyl)methacrylamide (HPMA) triblock copolymer. The polymer was synthesized and self-assembled into nanoparticles with 85 nm diameters. The copolymeric system with a high molecular weight (92 kDa) was degraded into small segments with molecular weights less than 50 kDa, which specifically released the anticancer drug paclitaxel in the cancer microenvironment. This type of system could lead to breast cancer treatment and MRI at the same time.

4.7 pH-RESPONSIVE NANOMATERIALS

The rapid proliferation of cancer cells causes glycolysis and lowers the pH in the tumor microenvironment, which aids in controlled drug release (Liberti and Locasale, 2016). When pH-sensitive nanoparticles are exposed to these acidic regions, their chemical structure changes, allowing the drug payload to be released (Gao et al., 2010). pH stimuli-responsive nanoparticles are widely made from both organic and inorganic materials (Gisbert-Garzarán et al., 2017). Dendritic polymers have been widely used in pH-sensitive systems due to their easy manipulation of solubility, conformation, and volume. The surface of dendritic polymers is treated with polyethylene glycol (PEG) to change their size, structure, and biocompatibility (Maeda, 2017). This type of polymer can improve drug loading and solubility while also increasing drug accumulation in tumor tissues due to increased permeability and retention. By forming a hydrazone bond between antitumor drug molecules and dendrimers, a more effective cancer treatment could be obtained (She et al., 2015).

Nanoparticles containing pH-sensitive precursor drugs capable of delivering hydrophobic anticancer drug combinations have been developed successfully (Zhang et al., 2016). PEG nanoparticles loaded with curcumin (CUR) and doxorubicin (DOX), for example, were combined with transferrin (Tf) to form nanocomplexes. Under mild acidic conditions, the simultaneous release of CUR and DOX was significantly accelerated. According to some studies, 79.2% and 57.6% of DOX is released from nanoparticles within 24 hours at pH values of 5.0 and 7.4, respectively (Cui et al., 2017). Because of their well-defined size and high drug encapsulation yields, Tf-attached nanocomplexes demonstrated advantages as drug-delivery carriers.

Self-assembled hyaluronic acid (HA) nanoparticles that use calcium phosphate to form hydroxyapatite nanoparticles loaded with DOX are one intriguing drug-delivery strategy. When this system is exposed to a low pH, the minerals in it dissolve, causing drug release at a specific tumor site (Han et al., 2013). Furthermore, pH-responsive AuNPs have been synthesized by introducing a mixed layer of single stranded DNA (ssDNA) and cytochrome c (Cyt c) on the surface of AuNPs by Park et al. (Figure 4.2). The synthesized Cyt c/ssDNA-AuNPs showed stability that was good enough to be used for next experiments in various buffer solutions as well as in deionized water. With those modified surfaces, in a normal physiological pH (~7.4), the surface charge of AuNPs is negative enough to repel each other due to electrostatic repulsions and the AuNPs exist individually within a cell. The charge of Cyt c on AuNPs surface, however, gradually changed to be positive with reduced pH (<6.5)

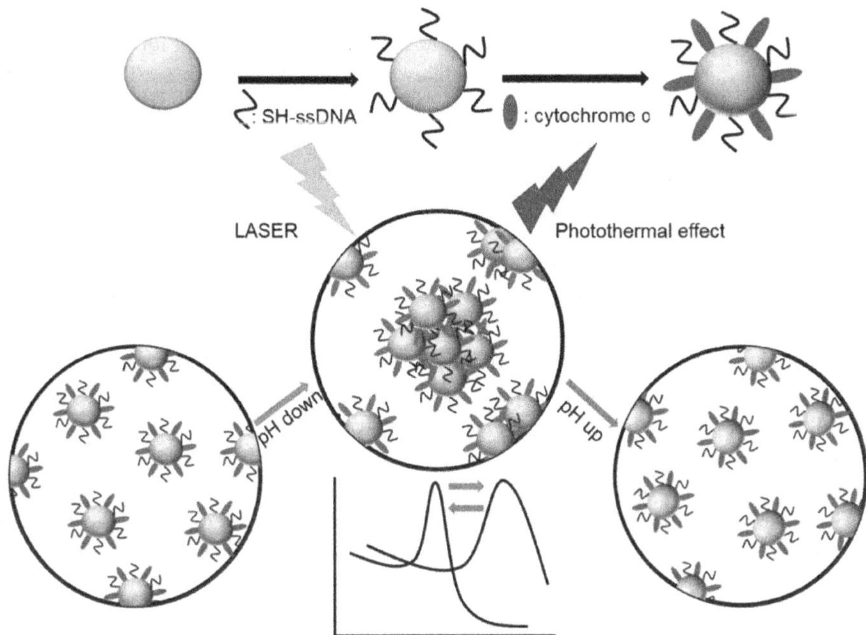

FIGURE 4.2 Schematic diagram of surface modification process of AuNPs with ssDNA and Cyt c modification process (upper) and pH-responsive behavior of Cyt c/ssDNA-AuNPs (lower). (Park et al., 2016.)

and eventually the AuNPs aggregated forming large clusters by local electrostatic attraction which is induced by the opposite charge of DNA strands and Cyt c.

4.8 BIOLOGICAL-RESPONSIVE NANOMATERIALS

Polymer nanomaterials, inorganic nanomaterials, metal-organic frameworks, and carbon-based nanomaterials are common examples of biologically responsive nanoparticles that have emerged as a result of significant advances in the materials science.

Organic nanoparticles have received a lot of attention in the biomedical field due to their inherent biocompatibility and biodegradability. Organic nanoparticles have high affinity and selectivity in detecting and binding with targeted biological entities such as proteins and peptides both *in vitro* and *in vivo*, making them a valuable biologically responsive nanomaterial. Polymeric nanoparticles capable of selectively recognizing targeted proteins or cells can also be created by immobilizing biologically responsive ligands on the particles' surface (e.g., antibodies, peptides, nucleic acids, and small molecules). In recent years, several ligand-modified nanoparticles have been developed to influence cell–cell, protein–protein, and protein–cell interactions. Liu et al., for example, used an A recognition element-modified nanoparticle to alter the morphology of A aggregates, resulting in the formation of co-assembled nanoclusters composed of A/nanoparticles rather than A oligomers. By reducing pathogenic A oligomers, this nanoparticle reduced A-induced neuron death (Zhao et al., 2018). Kim et al. created bio-specific multivalent nano-bioconjugate engager (dubbed mBiNE) using carboxylated polystyrene nanoparticles as a substrate, which could potentially boost immune-mediated tumor cell eradication. In recent decades, biologically sensitive inorganic nanoparticles (such as gold, silver, iron, platinum, titanium, cobalt, ceramic, and silica particles) have been used in regenerative medicine applications. A bone graft substitute, for example, should be capable of mimicking the extracellular matrix (ECM) of actual bone in order to provide excellent biocompatibility and sufficient mechanical strength for bone tissue regeneration. Because of their high mechanical strength, inorganic nanoparticles are a good candidate for bone graft replacement. Furthermore, inorganic nanoparticles can remain stable in the body for several weeks, assisting bone mending during the early stages of regeneration (Puppi et al., 2010). In bone tissue engineering, biologically responsive glasses, nanosilicates, hydroxyapatite, and silica nanoparticles have all been extensively used.

Recent advancements in Micro Ribonucleic Acid (miRNA) responsive systems use nanomaterials to deliver microRNA specifically to the target. MiRNAs are a type of small non-coding RNA that consists of short nucleotide sequences with 20–24 mer nucleotides that can act as key regulators of gene expression (Wery et al., 2011). Previous research has described the role of miRNA in the onset and progression of diseases such as cancer, cardiovascular disease, and others. Such research focuses on the differences in miRNA expression between normal and tumor regions, which has led to miRNA being considered as a therapeutic agent. Although there have been some advantages to using miRNA for therapeutic purposes, there are still some obstacles to miRNA delivery success. The limitations include the lack of cellular stability of miRNA, poor targeting ability, high toxicity, inability to deliver sufficient miRNAs to the desired tissues, and off-target effects of naked miRNA-based agents. Several strategies for improving miRNA stability have been investigated, including the use of ribose 2′-OH groups, locked nucleic acids, backbone modifications, and peptide nucleic acids (Bravo et al., 2007). However, the inefficient and non-specific delivery of miRNAs to cells remains a challenge.

Loading miRNA with nanoparticles that protect the miRNA from the external microenvironment has significantly improved the efficiency of delivering miRNA to the target, reducing degradation and increasing circulation time. Biocompatible nanocarriers with simple manufacturing requirements were used (Santos-Carballal et al., 2015).

Because of their biocompatibility, ease of synthesis, and tunable size and shape, gold nanoparticles (AuNPs) have received a lot of attention as nucleic acid delivery nanocarriers (Jain et al., 2006). To allow miRNA entrapment, the surface of AuNPs is modified with thiol or amino groups. Wang et al. created a composite of miRNA-124–5p and AuNPs to combine gene therapy and photothermal therapy for effective cancer cell killing. The cleavage of cystamine when AuNPs enter the cytoplasm of tumor cells via endocytosis could result in the release of miRNA-124–5p. Ekin et al. have also used AuNPs to deliver miRNA-145 into prostate and breast cancer cells (Ekin et al., 2014). The highly hydrophilic natural polysaccharide HA has been used to link polyethylenimine (PEI) and PEG to a controlled-release miRNA and/or DOX. Liang et al. created HA-NPs loaded with miRNA-145 to target colon cancer cells via their overexpressed CD44 receptors. Furthermore, the use of synthetic polymer nanoparticles for nucleic acid delivery has been studied. Liu et al. created PEG-peptide-polycaprolactone nanoparticles to deliver miRNA-200c and docetaxel. Wang et al. created PLGA-PEI NPs to deliver miRNA-542–3p and DOX to breast cancer cells (Wang et al., 2019). When compared to MCF-7 cells that express lower levels of CD44, a HA-decorated PEI-PLGA nanoparticle system increased both drug uptake and cytotoxicity in MDA-MB-231 cells (Ekin et al., 2014).

REFERENCES

Akhter, Md Habban, Habibullah Khalilullah, Manish Gupta, Mohamed A Alfaleh, Nabil A Alhakamy, Yassine Riadi, and Shadab Md. 2021. "Impact of protein corona on the biological identity of nanomedicine: Understanding the fate of nanomaterials in the biological milieu." *Biomedicines* no. 9 (10):1496.

Alromi, Dalal A, Seyed Yazdan Madani, and Alexander Seifalian. 2021. "Emerging application of magnetic nanoparticles for diagnosis and treatment of cancer." *Polymers* no. 13 (23):4146.

Bravo, Valia, Samuel Rosero, Camillo Ricordi, and Ricardo L Pastori. 2007. "Instability of miRNA and cDNAs derivatives in RNA preparations." *Biochemical and Biophysical Research Communications* no. 353 (4):1052–1055.

Cui, Tongxing, Sihao Zhang, and Hong Sun. 2017. "Co-delivery of doxorubicin and pH-sensitive curcumin prodrug by transferrin-targeted nanoparticles for breast cancer treatment." *Oncology Reports* no. 37 (2):1253–1260.

Ekin, Asli, Omer Faruk Karatas, Mustafa Culha, and Mustafa Ozen. 2014. "Designing a gold nanoparticle-based nanocarrier for microRNA transfection into the prostate and breast cancer cells." *The Journal of Gene Medicine* no. 16 (11–12):331–335.

Gao, Weiwei, Juliana M Chan, and Omid C Farokhzad. 2010. "pH-responsive nanoparticles for drug delivery." *Molecular Pharmaceutics* no. 7 (6):1913–1920.

Genchi, Giada Graziana, Attilio Marino, Agostina Grillone, Ilaria Pezzini, and Gianni Ciofani. 2017. "Remote control of cellular functions: The role of smart nanomaterials in the medicine of the future." *Advanced Healthcare Materials* no. 6 (9):1700002.

Gisbert-Garzarán, Miguel, Miguel Manzano, and María Vallet-Regí. 2017. "pH-responsive mesoporous silica and carbon nanoparticles for drug delivery." *Bioengineering* no. 4 (1):3.

Guardado-Alvarez, Tania M, Lekshmi Sudha Devi, Jean-Marie Vabre, Travis A Pecorelli, Benjamin J Schwartz, Jean-Olivier Durand, Olivier Mongin, Mireille Blanchard-Desce, and Jeffrey I Zink. 2014. "Photo-redox activated drug delivery systems operating under two photon excitation in the near-IR." *Nanoscale* no. 6 (9):4652–4658.

Gwon, Kihak, Eun-Jung Jo, Abhishek Sahu, Jae Young Lee, Min-Gon Kim, and Giyoong Tae. 2018. "Improved near infrared-mediated hydrogel formation using diacrylated Pluronic F127-coated upconversion nanoparticles." *Materials Science and Engineering: C* no. 90:77–84.

Han, Hwa Seung, Jungmin Lee, Hyun Ryoung Kim, Su Young Chae, Minwoo Kim, Gurusamy Saravanakumar, Hong Yeol Yoon, Dong Gil You, Hyewon Ko, and Kwangmeyung Kim. 2013. "Robust PEGylated hyaluronic acid nanoparticles as the carrier of doxorubicin: Mineralization and its effect on tumor targetability in vivo." *Journal of Controlled Release* no. 168 (2):105–114.

Huang, Chi, Zhengbo Sun, Haodong Cui, Ting Pan, Shengyong Geng, Wenhua Zhou, Paul K Chu, and Xue-Feng Yu. 2019. "InSe nanosheets for efficient NIR-II-responsive drug release." *ACS Applied Materials & Interfaces* no. 11 (31):27521–27528.

Jain, Prashant K, Kyeong Seok Lee, Ivan H El-Sayed, and Mostafa A El-Sayed. 2006. "Calculated absorption and scattering properties of gold nanoparticles of different size, shape, and composition: Applications in biological imaging and biomedicine." *The Journal of physical Chemistry B* no. 110 (14):7238–7248.

Khoee, Sepideh, and Mohammad Reza Karimi. 2018. "Dual-drug loaded Janus graphene oxide-based thermoresponsive nanoparticles for targeted therapy." *Polymer* no. 142:80–98.

Kim, Young-Jin, and Yukiko T Matsunaga. 2017. "Thermo-responsive polymers and their application as smart biomaterials." *Journal of Materials Chemistry B* no. 5 (23):4307–4321.

Kolosnjaj-Tabi, Jelena, Laure Gibot, Isabelle Fourquaux, Muriel Golzio, and Marie-Pierre Rols. 2019. "Electric field-responsive nanoparticles and electric fields: Physical, chemical, biological mechanisms and therapeutic prospects." *Advanced Drug Delivery Reviews* no. 138:56–67.

Lee, Hwangjae, Guru Karthikeyan Thirunavukkarasu, Semin Kim, and Jae Young Lee. 2018. "Remote induction of in situ hydrogelation in a deep tissue, using an alternating magnetic field and superparamagnetic nanoparticles." *Nano Research* no. 11 (11):5997–6009.

Li, Huipeng, Xue Yang, Zhanwei Zhou, Kaikai Wang, Chenzi Li, Hongzhi Qiao, David Oupicky, and Minjie Sun. 2017. "Near-infrared light-triggered drug release from a multiple lipid carrier complex using an all-in-one strategy." *Journal of Controlled Release* no. 261:126–137.

Li, Qingpo, Wei Li, Haixiao Di, Lihua Luo, Chunqi Zhu, Jie Yang, Xiaoyi Yin, Hang Yin, Jianqing Gao, and Yongzhong Du. 2018. "A photosensitive liposome with NIR light triggered doxorubicin release as a combined photodynamic-chemo therapy system." *Journal of Controlled Release* no. 277:114–125.

Liberti, Maria V, and Jason W Locasale. 2016. "The Warburg effect: How does it benefit cancer cells?" *Trends in Biochemical Sciences* no. 41 (3):211–218.

Liu, Qiang, Ji Chen, Yingru Li, and Gaoquan Shi. 2016. "High-performance strain sensors with fish-scale-like graphene-sensing layers for full-range detection of human motions." *ACS Nano* no. 10 (8):7901–7906.

Maeda, Hiroshi. 2017. "Polymer therapeutics and the EPR effect." *Journal of Drug Targeting* no. 25 (9–10):781–785.

Munaweera, Imalka, Sumbul Shaikh, Danny Maples, Adane S Nigatu, Sri Nandhini Sethuraman, Ashish Ranjan, David E Greenberg, and Rajiv Chopra. 2018. "Temperature-sensitive liposomal ciprofloxacin for the treatment of biofilm on infected metal implants using alternating magnetic fields." *International Journal of Hyperthermia* no. 34 (2):189–200.

Munaweera, Imalka, Yi Shi, Bhuvaneswari Koneru, Ruben Saez, Ali Aliev, Anthony J Di Pasqua, and Kenneth J Balkus Jr. 2015. "Chemoradiotherapeutic magnetic nanoparticles for targeted treatment of nonsmall cell lung cancer." *Molecular Pharmaceutics* no. 12 (10):3588–3596.

Neumann, S Ephraim, Christian F Chamberlayne, and Richard N Zare. 2018. "Electrically controlled drug release using pH-sensitive polymer films." *Nanoscale* no. 10 (21):10087–10093.

Park, Yeonju, Chihiro Hashimoto, Yukihiro Ozaki, and Young Mee Jung. 2016. "Understanding the phase transition of linear poly (N-isopropylacrylamide) gel under the heating and cooling processes." *Journal of Molecular Structure* no. 1124:144–150.

Puppi, Dario, Federica Chiellini, Anna Maria Piras, and Emo Chiellini. 2010. "Polymeric materials for bone and cartilage repair." *Progress in Polymer Science* no. 35 (4):403–440.

Sahle, Fitsum Feleke, Muhammad Gulfam, and Tao L Lowe. 2018. "Design strategies for physical-stimuli-responsive programmable nanotherapeutics." *Drug Discovery Today* no. 23 (5):992–1006.

Sánchez-Moreno, Paola, Juan De Vicente, Stefania Nardecchia, Juan A Marchal, and Houria Boulaiz. 2018. "Thermo-sensitive nanomaterials: Recent advance in synthesis and biomedical applications." *Nanomaterials* no. 8 (11):935.

Santos-Carballal, Beatriz, LJ Aaldering, Markus Ritzefeld, S Pereira, Norbert Sewald, BM Moerschbacher, Martin Götte, and Francisco Martin Goycoolea. 2015. "Physicochemical and biological characterization of chitosan-microRNA nanocomplexes for gene delivery to MCF-7 breast cancer cells." *Scientific Reports* no. 5 (1):1–15.

She, Wenchuan, Dayi Pan, Kui Luo, Bin He, Gang Cheng, Chengyuan Zhang, and Zhongwei Gu. 2015. "PEGylated dendrimer-doxorubicin cojugates as pH-sensitive drug delivery systems: Synthesis and in vitro characterization." *Journal of Biomedical Nanotechnology* no. 11 (6):964–978.

Shemetov, Anton A, Igor Nabiev, and Alyona Sukhanova. 2012. "Molecular interaction of proteins and peptides with nanoparticles." *ACS Nano* no. 6 (6):4585–4602.

Thirunavukkarasu, Guru Karthikeyan, Kondareddy Cherukula, Hwangjae Lee, Yong Yeon Jeong, In-Kyu Park, and Jae Young Lee. 2018. "Magnetic field-inducible drug-eluting nanoparticles for image-guided thermo-chemotherapy." *Biomaterials* no. 180:240–252.

Wang, Xiangdong, Nuo Jin, Qiao Wang, Tao Liu, Kangcan Liu, Yan Li, Yongkang Bai, and Xin Chen. 2019. "MiRNA delivery system based on stimuli-responsive gold nanoparticle aggregates for multimodal tumor therapy." *ACS Applied Bio Materials* no. 2 (7):2833–2839.

Wery, Maxime, Marta Kwapisz, and Antonin Morillon. 2011. "Noncoding RNAs in gene regulation." *Wiley Interdisciplinary Reviews: Systems Biology and Medicine* no. 3 (6):728–738.

Xiang, Jun, Xia Tong, Feng Shi, Qiang Yan, Bing Yu, and Yue Zhao. 2018. "Near-infrared light-triggered drug release from UV-responsive diblock copolymer-coated upconversion nanoparticles with high monodispersity." *Journal of Materials Chemistry B* no. 6 (21):3531–3540.

Yang, Guangbao, Jingjing Liu, Yifan Wu, Liangzhu Feng, and Zhuang Liu. 2016. "Near-infrared-light responsive nanoscale drug delivery systems for cancer treatment." *Coordination Chemistry Reviews* no. 320:100–117.

Yang, Hong Yu, Yi Li, and Doo Sung Lee. 2018a. "Multifunctional and stimuli-responsive magnetic nanoparticle-based delivery systems for biomedical applications." *Advanced Therapeutics* no. 1 (2):1800011.

Yang, Jun, Shaodong Zhai, Huan Qin, He Yan, Da Xing, and Xianglong Hu. 2018b. "NIR-controlled morphology transformation and pulsatile drug delivery based on multifunctional phototheranostic nanoparticles for photoacoustic imaging-guided photothermal-chemotherapy." *Biomaterials* no. 176:1–12.

Zhang, Yumin, Cuihong Yang, Weiwei Wang, Jinjian Liu, Qiang Liu, Fan Huang, Liping Chu, Honglin Gao, Chen Li, and Deling Kong. 2016. "Co-delivery of doxorubicin and curcumin by pH-sensitive prodrug nanoparticle for combination therapy of cancer." *Scientific Reports* no. 6 (1):1–12.

Zhao, Yu, Jinquan Cai, Zichen Liu, Yansheng Li, Chunxiong Zheng, Yadan Zheng, Qun Chen, Hongyun Chen, Feihe Ma, and Yingli An. 2018. "Nanocomposites inhibit the formation, mitigate the neurotoxicity, and facilitate the removal of β-amyloid aggregates in Alzheimer's disease mice." *Nano Letters* no. 19 (2):674–683.

5 SNM for Agricultural Applications

5.1 NANOFERTILIZERS

In order to enable controlled release and slow diffusion into the soil, nutrients are encapsulated or coated with nanomaterials to synthesize nanotechnology-based fertilizers. By minimizing nutrient loss through leaching and runoff and slowing down its rapid deterioration and volatility, nanoscale fertilizers can improve the soil fertility and the nutrient quality while also increasing the crop productivity over time (Nongbet et al., 2022). Moreover, nanofertilizers may be a good substitute for chemical fertilizers due to its high surface area to volume ratio properties and high penetration ability. Additionally, the use of nanofertilizers, also known as "nanobiofertilizers," can significantly lessen the environmental hazards (Bratovcic et al., 2021). As reported nanofertilizers may increase crop yield by promoting seed germination, nitrogen metabolism, photosynthesis, protein and carbohydrate synthesis, and stress tolerance (Rautela et al., 2021). In addition, relatively smaller amount of nanofertilizer need to be applied to the soil compared to the conventional fertilizers, which makes them less cost, easier to apply and transport.

However, using nanofertilizers has some restrictions and drawbacks, just like using other types of fertilizers (Zulfiqar et al., 2019). This chapter explains the use of nanofertilizers in sustainable and smart agriculture practices. The chapter will also cover the synthesis of nanofertilizers and the mechanistic explanation of how they increase soil fertility and increase crop yield, the benefits of nanofertilizers over traditional chemical fertilizers and restrictions for nanofertilizers.

Nanotechnology is the management and control of shape and size at the nanometer scale used in the synthesis and application of devices. It has paved the way and made it possible for "smart fertilizer," or the use of nanostructured materials as fertilizers (Wang et al., 2021). Additionally, nanofertilizers' chemical make-up can promote effective nutrient uptake, soil fertility restoration, ultra-high absorption, increased photosynthesis, increased production, decreased soil toxicity, decreased frequency of application, increased plant health, and decreased environmental pollution (Tyagi et al., 2022) (Figure 5.1). Silica, iron, zinc oxide, titanium dioxide, cerium oxide, aluminum oxide, gold nanorods, ZnCdSe/ZnS core-shell, InP/ZnS core-shell, and Mn/ZnSe quantum dots are examples of nanomaterial constituents in nanofertilizers (Basavegowda and Baek, 2021). The effectiveness of nanomaterials as nanofertilizers for plant growth depends significantly on their size, composition, concentration, and chemical properties, as well as the type of the crop (Figure 5.2). The nanofertilizers can be coated with polymers or thin coatings to encapsulate the NPs in order to prevent unfavorable nutrient losses (Rajonee et al., 2017).

DOI: 10.1201/9781003366270-5

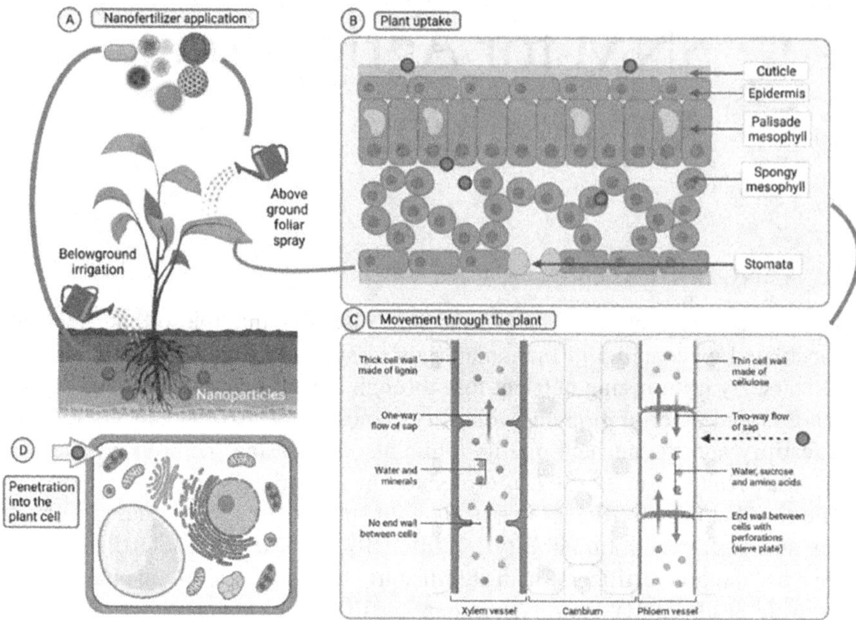

FIGURE 5.1 Application, uptake, translocation, and biodistribution of nanofertilizers inside the plant cells. (Nongbet et al., 2022.)

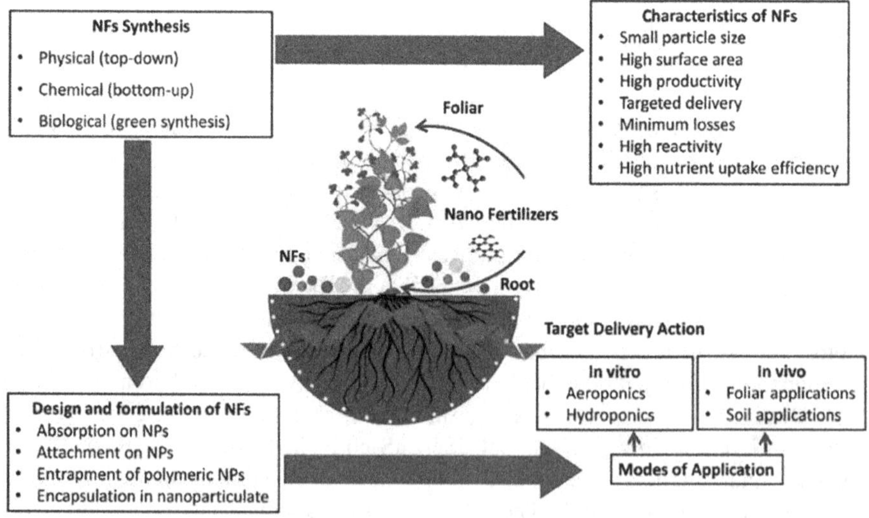

FIGURE 5.2 Overview of design and formulation of nanofertilizers based on the respective synthesis method and their characteristics. (Zahra et al., 2022.)

Application of nanofertilizers that makes the use of the special qualities of NPs can increase the effectiveness of nutrients use. Nanofertilizers can be created by adding nutrients singly or in combination to the adsorbents with the dimensions in nanoscale (Singh et al., 2021). The target nutrients are loaded, in the case of cationic nutrients, whereas the anionic nutrients are loaded following surface modification to produce the nanomaterials using physical and chemical techniques (Rajonee et al., 2016). The provision of nutrients as nanoscale particles or emulsions, their thin polymer coating, or their encapsulation within nanoporous materials are all examples of new technologies that allow for the encapsulation of fertilizers within NPs.

Some examples of nanofertilizers are mentioned in Table 5.1.

The effectiveness of the nanofertilizers is influenced by a number of factors. Factors that are intrinsic and extrinsic are heavily reliant on how nanofertilizers are absorbed, transported, accumulated, and exposed (Zulfiqar et al., 2019). Surface coatings and particle size are the main intrinsic factors that affect NPs efficacy, whereas soil texture, soil pH, and organic matter are the main extrinsic factors that have a significant impact on the potential use of NPs in crops. The behavior, bioavailability, and absorption of nanofertilizers in crops are significantly influenced by their

TABLE 5.1
Examples for Nanoferttilizers and Respective Crops Applied

Nanofertilizer	Applied Crops	References
Nanoencapsulated phosphorous	Maize	Naik et al. (2021)
Nano-chitosan-NPK fertilizers	Wheat	Abdel-Aziz et al. (2016)
Nano-chitosan	Pea	Okagu et al. (2021)
Nanopowder of cotton seed and ammonium fertilizer	Sweet potato	Iqbal (2019)
Aqueous solution on nanoiron	Cereals	Iqbal (2019)
Nanoparticles of ZnO	Cucumber, Peanut	Zhao et al. (2014a)
Rare-earth oxides nanoparticles	Vegetables	Iqbal (2019)
Nanosilver + allicin	Cereals	Abdelaal et al. (2020)
Iron oxide nanoparticles + calcium carbonate nanoparticles + peat	Cereals	Naik et al. (2021)
Sulfur nanoparticles + silicon dioxide nanoparticles + synthetic fertilizer	Cereals	Duhan et al. (2017)
Nano silicon dioxide	Maize	Kukreti et al. (2020)
Nano-TiO_2	Spinach	Zheng et al. (2005)
Gold nanoparticles + sulfur	Grapes	Malarkodi et al. (2017)
Nanocarbon + rare-earth metals + N fertilizers	Cereals	Sivarethinamohan and Sujatha (2021)
Stevia extract + nanoparticles of Se + organo-Ca + rare-earth elements + chitosan	Vegetables	Mastronardi et al. (2015)
Nanoiron slag powder	Maize	Sivarethinamohan and Sujatha (2021)
Nanoiron + organic manures	Cotton	Hussien et al. (2015)
Kaolin + SiO_2	Vegetables	Sivarethinamohan and Sujatha (2021)

uptake through the roots and leaves (El-Ramady et al., 2018). They slow and regulate the release mechanisms, which benefits farmers by regulating the accessibility of nutrients in the crops.

In comparison to conventional fertilizers, nanofertilizers are non-toxic and less harmful to both people and the environment. Additionally, they boost crop quality, yield, and soil fertility while cutting costs and maximizing the profit. According to a recent study by Carmona et al., post-synthesis and functionalization of amorphous calcium phosphate nanofertilizers, as opposed to their single-pot synthesis, significantly reduced manufacturing costs (Carmona et al., 2021). This new technique is now being used for the large-scale production of nanofertilizers in an effort to aid small-scale farmers and plant breeders. Chemical fertilizers are used extensively by farmers to increase crop production because they are synthetic, which means they are made of non-organic ingredients that are grown. Because they instantly dissolve in water, come in granular or liquid form, and are less expensive, chemical fertilizers work faster than organic fertilizers. However, some insoluble chemicals, such as mono-ammonium phosphate, diammonium phosphate, and triple superphosphate, are present in P fertilizers and do not readily dissolve in water (Nash and Halliwell, 1999). Soil microorganisms break down organic materials, such as animal manure, bird droppings, food scraps, and sewage sludge, releasing the necessary nutrients. This natural fertilizer is more environmentally friendly because it improves soil texture, boosts the activity of soil bacteria and fungi, and stores water for longer periods of time (Nongbet et al., 2022). The majority of the nutrients released from this material are nitrogen (N) and phosphorus (P). Since they enhance soil quality and contain the essential nutrients required for plant fertility and productivity, biofertilizers are great alternatives to synthetic fertilizers (Ali et al., 2021). They are also affordable, renewable, and considerate of the environment. *Pseudomonas* spp., which act as phosphate-solubilizing bacteria, and Azotobacter, Anabaena, and Rhizobium, which are involved in Nitrogen fixation, act as biofertilizers by assisting plants in nutrient uptake and absorption. In addition to fixing nitrogen and increasing the availability of nutrients for the plants, these microorganisms produce a variety of bioactive compounds, including organic acids, vitamins, growth hormones, and antagonistic compounds that protect plants from disease (Kalanaki et al., 2022).

5.2 NANOPESTICIDES/NANOHERBICIDES/NANOBACTERICIDES

Pesticides formulated in nanomaterials for agricultural applications, whether specially fixed on a hybrid substrate, encapsulated in a matrix, or functionalized nanocarriers for external stimuli or enzyme-mediated triggers, are referred to as nanopesticides (Rajna and Paschapur, 2019). Nanosized particles, combined with their shape and unique properties, are used as pesticides in nanocarrier innovative formulations based on a variety of materials such as silica, lipids, polymers, copolymers, ceramic, metal, carbon, and others (Agostini et al., 2012). Nanopesticide formulations can improve water solubility and bioavailability while also protecting agrochemicals from environmental degradation, revolutionizing crop pathogen, weed, and insect control. However, the nanomaterial properties are also on the verge of cytotoxicity and genotoxicity (Chaud et al., 2021) (Figure 5.3).

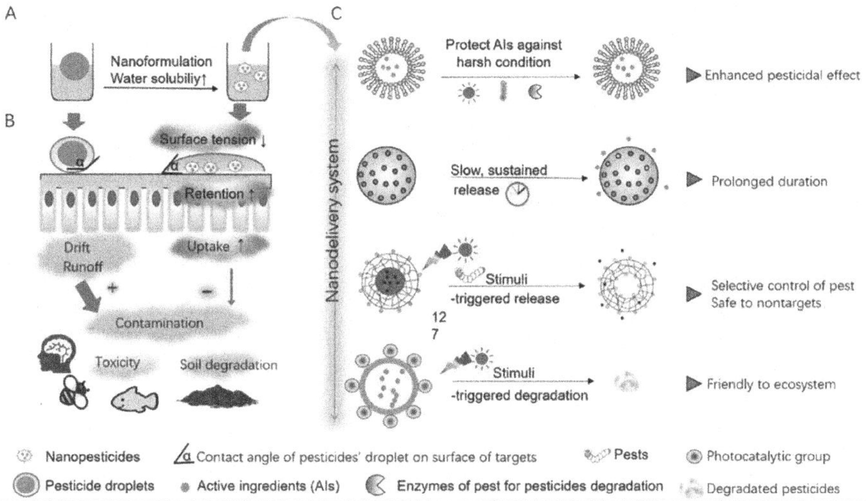

FIGURE 5.3 The benefits of nanotechnology for pest control. (a) Nanoformulation increases water solubility and stability of pesticidal active ingredients (AIs) by encapsulating AIs into small NCs-based pesticides that are conferring excellent characteristics at nanoscale. (b) Nanoformulation reduces surface tension of pesticidal droplets by decreasing the contact angle of droplets on the surface of targets (crop foliage), which contributes to a prolonged retention of droplets and a subsequently increased uptake ratio by target cells. This will be conducive to reduce the contamination caused by pesticides drift or runoff, bringing less toxicity to non-targets (mammal, pollinator, and aquatic organism) and bypassing yield reduction arising from the soil degradation. (c) Various nanodelivery systems that play a role in sustainable and smart application of pesticides. (Hou et al., 2021.)

Nanopesticides are formulated for their intended purpose, such as improving solubility, slowing the release of active ingredients, preventing degradation, and so on. To accomplish these goals, chemical nature carrier molecules have been modified and classified as organic polymer-based formulations, lipid-based formulations, nanosized metals and metal oxides, clay-based nanomaterials, and so on (Khan and Rizvi, 2017). This section discusses some of the most important nanoformulations that are being used as nanopesticides.

A nanoemulsion is an oil-in-water (O/W) emulsion in which the active ingredient of the chemical is dispersed as nanosized droplets in water, with surfactant molecules confined at the pesticide–water interface. Nano-emulsions are further classified as thermodynamically stable or kinetically stable based on the quantity and type of surfactants used. It is a thermodynamically stable nanoemulsion if the pesticide is partially soluble in the aqueous phase and spontaneous formation of a stable emulsion occurs when the surfactant, pesticide, and water components are combined (Rajna and Paschapur, 2019). Neem oil (*A. indica*) and citronella oil (*C. nardus*) are well known for their ability to repel pests. However, the low water solubility of these oils and other substances recognized as biopesticides restricts their use. The oils in the nanoemulsion system can be nanoencapsulated as a way to address this weakness. Neem oil and citronella oil were combined, and their effectiveness against fungi was

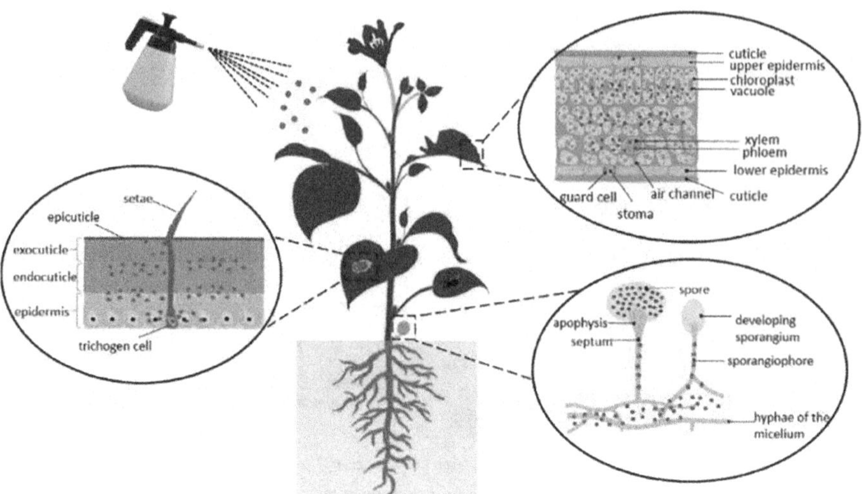

FIGURE 5.4 Schematic diagram showing the penetration of nanoemulsion-based agrochemicals through the leaves surface, insect cuticles and fungal body. (Mustafa and Hussein, 2020.)

evaluated, using the spontaneous emulsification technique (Chaud et al., 2021). A schematic diagram showing the penetration of nanoemulsion-based agrochemicals through the leaves surface, insect cuticles and fungal body is shown in Figure 5.4

The pesticide is dispersed as solid nanosized particles in aqueous media to create nano-suspensions, also known as nanodispersions. Surfactant molecules become confined at the particle surface in nanodispersions, with polar portions extending into the aqueous solution and non-polar portions associating with the solid pesticide (Chin et al., 2011). Pan et al. have synthesized nanosuspension of 5% lambda-cyhalothrin with 0.2% surfactants was prepared by the melt emulsification-high pressure homogenization method (Pan et al., 2015).

Polymer-based pesticide nanocarriers are widely used in the controlled and slow release of active ingredients to the target site. They can also be used to improve dispersion in aqueous media and as a protective reservoir. These products fall within the categories of nano-encapsulation, nano-spheres, nanogels, nano-fibers, etc. (Solanki et al., 2020).

Nanocapsules, also known as nano-encapsulation, are heterogeneous reservoir structures with an inner central cavity that confines the hydrophobic or hydrophilic active ingredient and is surrounded by a polymer coating or membrane. Pesticide-loaded nanocapsules are developed from polymers or during the polymerization of suitable monomers. Currently, there are many available nanocapsule formulations with several synthetic and natural polymers, such as polyethylene glycol (PEG), poly-ε-caprolactone (PCL), cellulose, chitosan, and alginate-gelatin (Yadav et al., 2021, Kumar et al., 2017).

Nanogels are made by cross-linking polymeric particles with hydrophilic groups, allowing them to absorb more water. Shen and co-workers have synthesized a series of poly(N-isopropylacrylamide) (PNIPAm)-based nanogels with high deformability and tuneable structure that were successfully constructed for smart pesticide delivery and effective pest control (Shen et al., 2022).

In addition to the formulations mentioned above, solid nanoparticles are tested to be used as nanopesticides (Hayles et al., 2017). Inert dusts such as silica, alumina, and clays cause sorption and abrasion damage to the wax coating on the insect cuticle. This physical damage can cause the insect to lose water, resulting in dehydration. Because of its additional benefit in increasing plant tolerance to abiotic and biotic stresses, nano-silica particles can be suggested as an eligible candidate among solid nanoparticles. However, high doses are required for a better result, which can have a negative impact on grain properties. Alumina nanostructures have also been reported to have pesticide properties. Moreover, nano-clays developed from montmorillonite were also shown to have pesticidal function but low toxicity.

Silver, titanium oxide, and copper are the most popular metal nanoparticles. Silver nanoparticles' bactericidal and virucidal activity makes them appealing to nanotechnology researchers. Metal NPs are more preferred in nanopesticide applications due to its low toxicity, inherent charge, larger surface area, and crystallographic structure (Rakowska et al., 2021). Titanium dioxide has been shown to have antimicrobial and antifungal activity in crops (Irshad et al., 2020). Nano-copper formulations have been shown to cause bacterial cell wall damage and to be effective against pomegranate bacterial blight at very low concentrations (Rai et al., 2018). Cell wall damage was observed in bacterial cells treated with nano-copper.

In general, targeted delivery and controlled release of nanopesticides can increase pesticide utilization while decreasing residue and pollution. Nano-microcapsule formulations, for example, have slow release and protection performance due to the use of light-sensitive, thermo-sensitive, humidity-sensitive, enzyme-sensitive, and soil pH-sensitive high polymer materials to deliver pesticides. The dispersion and bioactivity of the active component of pesticide molecules are improved by nanopesticide formulations, which also reduce drift losses by increasing droplet adhesion to plant surfaces. Because of this, nanopesticides will have a higher efficacy than conventional pesticide formulations (such as Dust, Granule, Pellets, Emulsifiable Concentrate, Wettable Powder, Water Dispersible Granule, etc.) (Hazra and Purkait, 2019), and because of their small size, improved pesticide droplet ductility, wettability, and target adsorption when sprayed in fields.

5.3 NANOSENSORS

Nanosensors have proven their outstanding performance in the field of agriculture. Moreover, nanosensors enable real-time crop and field conditions, crop growth, pest activity, plant diseases, and environmental stressors easier to monitor (Chen and Yada, 2011). The fabrication of such novel nanosensors has played a significant role in the development of agricultural fields. The real-time monitoring in the agricultural crops and environment conditions have prevented the excess use of pesticides and fertilizer amounts, which is helpful in the reduction of environmental contamination and product cost (Chhipa, 2019). Nanosensor applications transform traditional agricultural methods into smart agriculture, which uses more environmentally friendly and energy-efficient methods for sustainable agricultural practices. According to Fraceto et al. (2016), smart agriculture practices included three factors; fertilizer or pesticide delivery systems based on nanoformulations that increase nutrient dispersion and wettability, a nanodetector for pesticide or fertilizer residues, and finally remote-sensing-based monitor systems for disease occurrence and crop growth (Fraceto et al., 2016). In agriculture, nanosensors are used to measure soil moisture, pesticide residue, nutrient needs, and crop pest identification (Kaushal and Wani, 2017). The high sensitivity and detection limit of nanosensors make them more advantageous for smart agricultural applications (Dhiman et al., 2021). The development of nanosensors for fertilizer detection has made use of a variety of metal nanomaterials, such as gold nanoparticles (AuNPs), carbon nanotubes (CNT), quantum dots (QD), and various nanocomposites with polymers (Omanović-Mikličanina and Maksimović, 2016).

The nanosensors outperformed conventional sensors in many ways, including high sensitivity due to a high surface to volume ratio, a rapid response time of only a few seconds, more consistent and reliable results, the ability to detect small amounts of matter (up to a nanogram or less), the ability to be used in a variety of matrices, and support for rapid electron transfer kinetics (Adam and Gopinath, 2022). Some examples for the currently used nanosensors in agricultural field are mentioned in Table 5.2.

TiO_2 nanotubes based on nanosensors were created by Yu et al. (2010) for the detection of atrazine in soil at levels of parts per trillion (ppt) (Yu et al., 2010). Furthermore, by Chen et al., synthesized chitosan nanocomposites modified with

TABLE 5.2
Examples of Nanosensors and Their Targeted Compounds

Nanomaterial in Nanosensor	Target Compound/Species	References
Titanium oxide (TiO_2)	Atrazine	Yu et al. (2010)
Graphene	Herbicide detection	Zhao et al. (2011)
Carbon	Herbicide detection	Luo et al. (2014)
Gold (Au)	Acetamiprid	Shi et al. (2013)
Gold (Au)	Urea and urease activity	Deng et al. (2016)
Gold (Au)	*Pantoea stewartii* sbusp.	Zhao et al. (2014b)
Gold (Au)	Herbicide detection	Boro et al. (2011)
Gold (Au)	Organophosphates detection	Kang et al. (2010)
Gold nanorods	Cymbidium mosaic virus and Odontoglossum ringspot virus	Lin et al. (2014)
Graphene oxide	Nitrate	Radhakrishnan et al. (2014)
Multi-walled Carbon nanotubes (MWCNT)	Glyphosate and glufosinate	Chhipa (2019)
MWC-chitosan nanocomposite	Methyl parathion	Zhao et al. (2021)
Quantum dots	DNA	Mohd Bakhori et al. (2013)
Green fluorescent proteins, also known as "chameleons"	Ca^{2+}	Truong et al. (2001)
Silver (Ag)	Herbicide detection	Dubas and Pimpan (2008)
ZnO-chitosan nanocomposite membrane	Trichoderma harzianum	Siddiquee et al. (2014)

glassy carbon electrode and immobilized acetylcholinesterase enzyme on multi-walled carbon nanotubes (MWCNT). This sensor was used to detect methyl parathion at ultra-trace levels in soil and water (Chen et al., 2015). Based on the acetylcholinesterase enzyme's inhibitory effect, methyl parathion was detected. Similarly, the development of a nanobiosensor based on AuNPs functionalized with an aptamer that binds to acetamiprid allowed for the measurement of acetamiprid detection in soil. Acetamiprid can be seen being detected by this nanobiosensor at concentrations between 75 nM and 7.5 M (Tseng et al., 2020).

The continuous detection of soil conditioning needs requires a quick estimation of nutrient concentrations in soil, which is a major problem. Nowadays, the field of developing nanobiosensors for fertilizer estimation is expanding. The sensors based on nanotechnology can deliver precise data on fertilizer requirements, which can be useful in lowering costs for farmers and saving unused fertilizers. In-depth reviews on the use of nanosensors in agriculture have been published in previous reports (Antonacci et al., 2018). In addition, Mura et al. (2015) developed a colorimetric assay to find nitrate in soil using AuNPs modified with cysteamine (Mura et al., 2015). Similar to this, Li and co-workers created graphene oxide-based nanosensors for the detection of nitrate (Li et al., 2019), while Azahar Ali et al. (2017) created a nitrate detection sensor using poly(3,4-ethylenedioxythiophene) nanofibres and graphite oxide nanosheets (Ali et al., 2017). The creation of a biosensor based on the AuNP-3,3′,5,5′-tetramethylbenzidine-H_2O_2 reaction could enable the detection of urea, urease activity, and urease inhibition (Deng et al., 2016). In this system, AuNPs serve as a catalyst and the reaction results in a yellow color. The detection limit for urease activity in soil is 1.8 U/L.

On the other hand, pest detection is a crucial component of the agricultural field. Traditional detection techniques take a lot of time and are not efficient. Crops can be protected and damage can be controlled by the rapid and accurate results produced by the use of nanosensors for pest detection. A quantum dot-FRET (Fluorescence Resonance Energy Transfer) based nanobiosensor was created by Safarpour et al. to detect Polymyxa betae (Safarpour et al., 2012). One of the most important significance of the combination of QDs and FRET systems is their application as the QD-FRET based biosensors in designing biomolecule detection systems specially used for pathogen detection in the samples of interest. Furthermore, Bakhori et al. used a synthetic oligonucleotide in conjunction with a FRET system to identify Ganoderma boninense (Mohd Bakhori et al., 2013). Additionally, DNA probes and modified quantum dots were used to create the sensor. Pathogen detection has also been done with the aid of AuNPs. Furthermore, Siddiquee et al. (2014) created a nanobiosensor using a modified gold electrode and a ZnO nanoparticle/chitosan nanocomposite to detect the soil-borne fungal pathogen *Trichoderma harzianum* (Siddiquee et al., 2014). In order to find the bacterial plant pathogen Xanthomonas axonopodis pv. vesicatoria in Solanaceous crops, Yao et al. (2009) used fluorescent silica nanoprobes conjugated with a secondary antibody of goat anti rabbit Ig (Yao et al., 2009).

Novel nanosensors that allow the real-time demonstration of significant intracellular or extracellular factors are also being developed in addition to biological nanosensors. The vital element determining the activity of many biological reactions is the hydrogen concentration, or pH, in the cell or tissue. Uchiyama and Makino showed how to incorporate a water-sensitive fluorophore into a pH-responsive polymer to create digital-type fluorescent pH sensors (Uchiyama and Makino, 2009). It is noteworthy that a large number of digital signal transductions involve biological macromolecules as the digital processors in living things.

A specific subcellular compartment containing intracellular redox-active labile iron was discovered using iron-sensitive fluorescent chemosensors and digital fluorescence spectroscopy (Fakih et al. 2008), Calcium is another crucial ion in a cell's system. This ion plays a crucial role in the intracellular signaling processes and aids in the homeostasis of mineralized tissues. A common protein called calmodulin (CaM) is involved in Ca^{2+}-mediated signal transduction. CaM develops a strong affinity for a variety of cellular proteins with one or more CaM recognition sequences upon Ca^{2+} influx, causing the start or stoppage of Ca^{2+}-regulated cascades. Green fluorescent proteins, also known as "chameleons," and CaM complexes were used by Truong et al. to create protein-based Ca^{2+} sensors (Truong et al., 2001). The main benefit of chameleons is their ability to measure localized Ca^{2+} changes because they can be expressed in single cells and directed to particular organelles or tissues.

REFERENCES

Abdel-Aziz, Heba MM, Mohammed NA Hasaneen, and Aya M Omer. 2016. "Nano chitosan-NPK fertilizer enhances the growth and productivity of wheat plants grown in sandy soil." *Spanish Journal of Agricultural Research* no. 14 (1):e0902–e0902.

Abdelaal, Khaled AA, El A EL-Shawy, Yaser Mohamed Hafez, SM Abdel-Dayem, Russel Chrispine Garven Chidya, Hirofumi Saneoka, and Ayman El Sabagh. 2020. "Nano-Silver and non-traditional compounds mitigate the adverse effects of net blotch disease of barley in correlation with up-regulation of antioxidant enzymes." *Pakistan Journal of Botany* no. 52 (3):1065–1072.

Adam, Tijjani, and Subash CB Gopinath. 2022. "Nanosensors: Recent perspectives on attainments and future promise of downstream applications." *Process Biochemistry* no. 117:153–173.

Agostini, Alessandro, Laura Mondragón, Carmen Coll, Elena Aznar, M Dolores Marcos, Ramón Martínez-Máñez, Félix Sancenón, Juan Soto, Enrique Pérez-Payá, and Pedro Amorós. 2012. "Dual enzyme-triggered controlled release on capped nanometric silica mesoporous supports." *ChemistryOpen* no. 1 (1):17–20.

Ali, Md Azahar, Huawei Jiang, Navreet K Mahal, Robert J Weber, Ratnesh Kumar, Michael J Castellano, and Liang Dong. 2017. "Microfluidic impedimetric sensor for soil nitrate detection using graphene oxide and conductive nanofibers enabled sensing interface." *Sensors and Actuators B: Chemical* no. 239:1289–1299.

Ali, Sameh S, Rania Al-Tohamy, Eleni Koutra, Mohamed S Moawad, Michael Kornaros, Ahmed M Mustafa, Yehia A-G Mahmoud, Abdelfattah Badr, Mohamed EH Osman, and Tamer Elsamahy. 2021. "Nanobiotechnological advancements in agriculture and food industry: Applications, nanotoxicity, and future perspectives." *Science of the Total Environment* no. 792:148359.

Antonacci, Amina, Fabiana Arduini, Danila Moscone, Giuseppe Palleschi, and Viviana Scognamiglio. 2018. "Nanostructured (bio) sensors for smart agriculture." *TrAC Trends in Analytical Chemistry* no. 98:95–103.

Basavegowda, Nagaraj, and Kwang-Hyun Baek. 2021. "Current and future perspectives on the use of nanofertilizers for sustainable agriculture: The case of phosphorus nanofertilizer." *3 Biotech* no. 11 (7):1–21.

Boro, Robin Chandra, Jyotsna Kaushal, Yogesh Nangia, Nishima Wangoo, Aman Bhasin, and C Raman Suri. 2011. "Gold nanoparticles catalyzed chemiluminescence immunoassay for detection of herbicide 2, 4-dichlorophenoxyacetic acid." *Analyst* no. 136 (10):2125–2130.

Bratovcic, Amra, Wafaa M Hikal, Hussein AH Said-Al Ahl, Kirill G Tkachenko, Rowida S Baeshen, Ali S Sabra, and Hoda Sany. 2021. "Nanopesticides and nanofertilizers and agricultural development: Scopes, advances and applications." *Open Journal of Ecology* no. 11 (4):301–316.

Carmona, Francisco J, Gregorio Dal Sasso, Gloria B Ramírez-Rodríguez, Youry Pii, José Manuel Delgado-López, Antonietta Guagliardi, and Norberto Masciocchi. 2021. "Urea-functionalized amorphous calcium phosphate nanofertilizers: Optimizing the synthetic strategy towards environmental sustainability and manufacturing costs." *Scientific Reports* no. 11 (1):1–14.

Chaud, Marco, Eliana B Souto, Aleksandra Zielinska, Patricia Severino, Fernando Batain, Jose Oliveira-Junior, and Thais Alves. 2021. "Nanopesticides in agriculture: Benefits and challenge in agricultural productivity, toxicological risks to human health and environment." *Toxics* no. 9 (6):131.

Chen, Dongfei, Yancui Jiao, Huiying Jia, Yemin Guo, Xia Sun, Xiangyou Wang, and Jianguang Xu. 2015. "Acetylcholinesterase biosensor for chlorpyrifos detection based on multi-walled carbon nanotubes-SnO_2-chitosan nanocomposite modified screen-printed electrode." *International Journal of Electrochemical Science* no. 10:10491–10501.

Chen, Hongda, and Rickey Yada. 2011. "Nanotechnologies in agriculture: New tools for sustainable development." *Trends in Food Science & Technology* no. 22 (11):585–594.

Chhipa, Hemraj. 2019. "Applications of nanotechnology in agriculture." In Volker Gurtler, Andrew S. Ball, and Sarvesh Soni (Eds.), *Methods in microbiology*, 115–142. Elsevier, Amsterdam.

Chin, Chih-Ping, Ho-Shing Wu, and Shaw S Wang. 2011. "New approach to pesticide delivery using nanosuspensions: Research and applications." *Industrial & Engineering Chemistry Research* no. 50 (12):7637–7643.

Deng, Hao-Hua, Guo-Lin Hong, Feng-Lin Lin, Ai-Lin Liu, Xing-Hua Xia, and Wei Chen. 2016. "Colorimetric detection of urea, urease, and urease inhibitor based on the peroxidase-like activity of gold nanoparticles." *Analytica Chimica Acta* no. 915:74–80.

Dhiman, Shikha, Annu Yadav, Nitai Debnath, and Sumistha Das. 2021. "Application of core/shell nanoparticles in smart farming: A paradigm shift for making the agriculture sector more sustainable." *Journal of Agricultural and Food Chemistry* no. 69 (11):3267–3283.

Dubas, Stephan T, and Vimolvan Pimpan. 2008. "Humic acid assisted synthesis of silver nanoparticles and its application to herbicide detection." *Materials Letters* no. 62 (17–18):2661–2663.

Duhan, Joginder Singh, Ravinder Kumar, Naresh Kumar, Pawan Kaur, Kiran Nehra, and Surekha Duhan. 2017. "Nanotechnology: The new perspective in precision agriculture." *Biotechnology Reports* no. 15:11–23.

El-Ramady, Hassan, Neama Abdalla, Tarek Alshaal, Ahmed El-Henawy, Mohammed Elmahrouk, Yousry Bayoumi, Tarek Shalaby, Megahed Amer, Said Shehata, and Miklós Fári. 2018. "Plant nano-nutrition: Perspectives and challenges." In K M Gothandam, Shivendu Ranjan, Nandita Dasgupta, Chidambaram Ramalingam, Eric Lichtfouse (Eds.), *Nanotechnology, food security and water treatment*, 129–161, Springer Cham, New York City.

Fakih, Sarah, Maria Podinovskaia, Xiaole Kong, Helen L Collins, Ulrich E Schaible, and Robert C Hider. 2008. "Targeting the lysosome: Fluorescent iron (III) chelators to selectively monitor endosomal/lysosomal labile iron pools." *Journal of Medicinal Chemistry* no. 51 (15):4539–4552.

Fraceto, Leonardo F, Renato Grillo, Gerson A de Medeiros, Viviana Scognamiglio, Giuseppina Rea, and Cecilia Bartolucci. 2016. "Nanotechnology in agriculture: Which innovation potential does it have?" *Frontiers in Environmental Science* no. 4:20.

Hayles, John, Lucas Johnson, Clare Worthley, and Dusan Losic. 2017. "Nanopesticides: A review of current research and perspectives." In Alexandru Mihai Grumezescu (Ed.), *New pesticides and soil sensors*, 193–225, Academic press, Cambridge, MA.

Hazra, Dipak Kumar, and Aloke Purkait. 2019. "Role of pesticide formulations for sustainable crop protection and environment management: A review." *Journal of Pharmacognosy and Phytochemistry* no. 8:686–693.

Hou, Qiuli, Hanqiao Zhang, Lixia Bao, Zeyu Song, Changpeng Liu, Zhenqi Jiang, and Yang Zheng. 2021. "NCs-delivered pesticides: A promising candidate in smart agriculture." *International Journal of Molecular Sciences* no. 22 (23):13043.

Hussien, MM, Soad M El-Ashry, Wafaa M Haggag, and Dalia M Mubarak. 2015. "Response of mineral status to nano-fertilizer and moisture stress during different growth stages of cotton plants." *International Journal of ChemTech Research* no. 8:643–650.

Iqbal, Muhammad Aamir. 2019. "Nano-fertilizers for sustainable crop production under changing climate: A global perspective." *Sustainable Crop Production* no. 8:1–13.

Irshad, Muhammad Atif, Rab Nawaz, Muhammad Zia ur Rehman, Muhammad Imran, Jamil Ahmad, Sajjad Ahmad, Aqil Inam, Abdul Razzaq, Muhammad Rizwan, and Shafaqat Ali. 2020. "Synthesis and characterization of titanium dioxide nanoparticles by chemical and green methods and their antifungal activities against wheat rust." *Chemosphere* no. 258:127352.

Kalanaki, Mahdi, Fatemeh Karandish, Payman Afrasiab, Henk Ritzema, Issa Khamari, and Seyed Mahmood Tabatabai. 2022. "Assessing the influence of integrating soil amendment applications with saline water irrigation on Ajwain's yield and water productivity." *Irrigation Science* no. 40 (1):71–85.

Kang, Tian-Fang, Fei Wang, Li-Ping Lu, Yan Zhang, and Tong-Shen Liu. 2010. "Methyl parathion sensors based on gold nanoparticles and Nafion film modified glassy carbon electrodes." *Sensors and Actuators B: Chemical* no. 145 (1):104–109.

Kaushal, Manoj, and Suhas P Wani. 2017. "Nanosensors: Frontiers in precision agriculture." In Ram Prasad, Manoj Kumar, and Vivek Kumar (Eds.), *Nanotechnology,* 279–291. Springer, New York City.

Khan, Mujeebur Rahman, and Tanveer Fatima Rizvi. 2017. "Application of nanofertilizer and nanopesticides for improvements in crop production and protection." In Mansour Ghorbanpour, Khanuja Manika, and Ajit Varma (Eds.), *Nanoscience and plant–soil systems*, 405–427. Springer, New York City.

Kukreti, Bharti, Anita Sharma, Parul Chaudhary, Upasana Agri, and Damini Maithani. 2020. "Influence of nanosilicon dioxide along with bioinoculants on Zea mays and its rhizospheric soil." *3 Biotech* no. 10 (8):1–11.

Kumar, Sandeep, Gaurav Bhanjana, Amit Sharma, Neeraj Dilbaghi, MC Sidhu, and Ki-Hyun Kim. 2017. "Development of nanoformulation approaches for the control of weeds." *Science of the Total Environment* no. 586:1272–1278.

Li, Daoliang, Tan Wang, Zhen Li, Xianbao Xu, Cong Wang, and Yanqing Duan. 2019. "Application of graphene-based materials for detection of nitrate and nitrite in water—a review." *Sensors* no. 20 (1):54.

Lin, Hsing-Ying, Chen-Han Huang, Sin-Hong Lu, I-Ting Kuo, and Lai-Kwan Chau. 2014. "Direct detection of orchid viruses using nanorod-based fiber optic particle plasmon resonance immunosensor." *Biosensors and Bioelectronics* no. 51:371–378.

Luo, Mai, Donghui Liu, Lu Zhao, Jiajun Han, Yiran Liang, Peng Wang, and Zhiqiang Zhou. 2014. "A novel magnetic ionic liquid modified carbon nanotube for the simultaneous determination of aryloxyphenoxy-propionate herbicides and their metabolites in water." *Analytica Chimica Acta* no. 852:88–96.

Malarkodi, Chelladurai, Shanmugam Rajeshkumar, and Gurusamy Annadurai. 2017. "Detection of environmentally hazardous pesticide in fruit and vegetable samples using gold nanoparticles." *Food Control* no. 80:11–18.

Mastronardi, Emily, Phepafatso Tsae, Xueru Zhang, Carlos Monreal, and Maria C DeRosa. 2015. "Strategic role of nanotechnology in fertilizers: Potential and limitations." In Mahendra Rai, Caue Ribeiro, Luiz Mattoso, and Nelson Duran (Eds.), *Nanotechnologies in food and agriculture*, 25–67. Springer, New York City.

Mohd Bakhori, Noremylia, Nor Azah Yusof, Abdul Halim Abdullah, and Mohd Zobir Hussein. 2013. "Development of a fluorescence resonance energy transfer (FRET)-based DNA biosensor for detection of synthetic oligonucleotide of Ganoderma boninense." *Biosensors* no. 3 (4):419–428.

Mura, Stefania, Gianfranco Greppi, Pier Paolo Roggero, Elodia Musu, Daniele Pittalis, Alberto Carletti, Giorgio Ghiglieri, and Joseph Irudayaraj. 2015. "Functionalized gold nanoparticles for the detection of nitrates in water." *International Journal of Environmental Science and Technology* no. 12 (3):1021–1028.

Mustafa, Isshadiba Faikah, and Mohd Zobir Hussein. 2020. "Synthesis and technology of nanoemulsion-based pesticide formulation." *Nanomaterials* no. 10 (8):1608.

Naik, B Sri Sai Siddartha, Neeta Mahawar, Tirunagari Rupesh, Swetha Dhegavath, and Raghuvir Singh Meena. 2021. "Nano-technology based nano-fertilizer: A sustainable approach for enhancing crop productivity under climate changing situations." *Current Research in Agriculture and Farming* no. 2 (1):21–29.

Nash, David M, and David J Halliwell. 1999. "Fertilisers and phosphorus loss from productive grazing systems." *Soil Research* no. 37 (3):403–430.

Nongbet, Amilia, Awdhesh Kumar Mishra, Yugal Kishore Mohanta, Saurov Mahanta, Manjit Kumar Ray, Maryam Khan, Kwang-Hyun Baek, and Ishani Chakrabartty. 2022. "Nanofertilizers: A smart and sustainable attribute to modern agriculture." *Plants* no. 11 (19):2587.

Okagu, Ogadimma D, Jian Jin, and Chibuike C Udenigwe. 2021. "Impact of succinylation on pea protein-curcumin interaction, polyelectrolyte complexation with chitosan, and gastrointestinal release of curcumin in loaded-biopolymer nano-complexes." *Journal of Molecular Liquids* no. 325:115248.

Omanović-Mikličanina, Enisa, and Mirjana Maksimović. 2016. "Nanosensors applications in agriculture and food industry." *Bulletin of the Chemists and Technologists of Bosnia and Herzegovina* no. 47:59–70.

Pan, Zhenzhong, Bo Cui, Zhanghua Zeng, Lei Feng, Guoqiang Liu, Haixin Cui, and Hongyu Pan. 2015. "Lambda-cyhalothrin nanosuspension prepared by the melt emulsification-high pressure homogenization method." *Journal of Nanomaterials* no. 2015:263.

Radhakrishnan, Sivaprakasam, Karthikeyan Krishnamoorthy, Chinnathambi Sekar, Jeyaraj Wilson, and Sang Jae Kim. 2014. "A highly sensitive electrochemical sensor for nitrite detection based on Fe_2O_3 nanoparticles decorated reduced graphene oxide nanosheets." *Applied Catalysis B: Environmental* no. 148:22–28.

Rai, Mahendra, Avinash P Ingle, Raksha Pandit, Priti Paralikar, Sudhir Shende, Indarchand Gupta, Jayanta K Biswas, and Silvio Silvério da Silva. 2018. "Copper and copper nanoparticles: Role in management of insect-pests and pathogenic microbes." *Nanotechnology Reviews* no. 7 (4):303–315.

Rajna, S, and AU Paschapur. 2019. "Nanopesticides: Its scope and utility in pest management." *Indian Farmer* no. 6 (1):17–21.

Rajonee, Anjuman Ara, Farah Nigar, Samina Ahmed, and SM Imamul Huq. 2016. "Synthesis of nitrogen nano fertilizer and its efficacy." *Canadian Journal of Pure and Applied Sciences* no. 10:3913–3919.

Rajonee, Anjuman Ara, Shurovi Zaman, and Shah Muhammad Imamul Huq. 2017. "Preparation, characterization and evaluation of efficacy of phosphorus and potassium incorporated nano fertilizer." *Advances in Nanoparticles* no. 6 (2):62.

Rakowska, Paulina D, Mariavitalia Tiddia, Nilofar Faruqui, Claire Bankier, Yiwen Pei, Andrew J Pollard, Junting Zhang, and Ian S Gilmore. 2021. "Antiviral surfaces and coatings and their mechanisms of action." *Communications Materials* no. 2 (1):1–19.

Rautela, Indra, Pallavi Dheer, Priya Thapliyal, Dheeraj Shah, Mallika Joshi, Shuchi Upadhyay, Prateek Gururani, Vimlendu Bhushan Sinha, Naveen Gaurav, and Manish Dev Sharma. 2021. "Current scenario and future perspectives of nanotechnology in sustainable agriculture and food production." *Plant Cell Biotechnology and Molecular Biology* no. 22 (11&12):99–121.

Safarpour, Hossein, Mohammad Reza Safarnejad, Meisam Tabatabaei, Afshin Mohsenifar, Fatemeh Rad, Marzieh Basirat, Fatemeh Shahryari, and Fatemeh Hasanzadeh. 2012. "Development of a quantum dots FRET-based biosensor for efficient detection of Polymyxa betae." *Canadian Journal of Plant Pathology* no. 34 (4):507–515.

Shen, Yue, Changcheng An, Jiajun Jiang, Bingna Huang, Ningjun Li, Changjiao Sun, Chong Wang, Shenshan Zhan, Xingye Li, and Fei Gao. 2022. "Temperature-dependent nanogel for pesticide smart delivery with improved foliar dispersion and bioactivity for efficient control of multiple pests." *ACS Nano* no. 16:20622–20632.

Shi, Huijie, Guohua Zhao, Meichuan Liu, Lifang Fan, and Tongcheng Cao. 2013. "Aptamer-based colorimetric sensing of acetamiprid in soil samples: Sensitivity, selectivity and mechanism." *Journal of Hazardous Materials* no. 260:754–761.

Siddiquee, Shafiquzzaman, Kobun Rovina, Nor Azah Yusof, Kenneth Francis Rodrigues, and Saallah Suryani. 2014. "Nanoparticle-enhanced electrochemical biosensor with DNA immobilization and hybridization of *Trichoderma harzianum* gene." *Sensing and Bio-Sensing Research* no. 2:16–22.

Singh, Harpreet, Archita Sharma, Sanjeev K Bhardwaj, Shailendra Kumar Arya, Neha Bhardwaj, and Madhu Khatri. 2021. "Recent advances in the applications of nano-agrochemicals for sustainable agricultural development." *Environmental Science: Processes & Impacts* no. 23 (2):213–239.

Sivarethinamohan, R, and S Sujatha. 2021. Unlocking the potentials of using nanotechnology to stabilize agriculture and food production. Paper read at AIP Conference Proceedings.

Solanki, Chandresh B, V Birari Vaishali, Manishkumar J Joshi, and V Prithiv Raj. 2020 "Nenopesticide: Its role in pest management." *Agriculture and Food: E-Newsletter* no. 2 (10), 32125.

Truong, Kevin, Asako Sawano, Hideaki Mizuno, Hiroshi Hama, Kit I Tong, Tapas Kumar Mal, Atsushi Miyawaki, and Mitsuhiko Ikura. 2001. "FRET-based in vivo Ca^{2+} imaging by a new calmodulin-GFP fusion molecule." *Nature Structural Biology* no. 8 (12):1069–1073.

Tseng, Wei-Bin, Ming-Mu Hsieh, Che-Hsie Chen, Tai-Chia Chiu, and Wei-Lung Tseng. 2020. "Functionalized gold nanoparticles for sensing of pesticides: A review." *Journal of Food & Drug Analysis* no. 28 (4):521.

Tyagi, Jyoti, Shahzad Ahmad, and Moksh Malik. 2022. "Nitrogenous fertilizers: Impact on environment sustainability, mitigation strategies, and challenges." *International Journal of Environmental Science and Technology* no. 11:11649–11672.

Uchiyama, Seiichi, and Yumi Makino. 2009. "Digital fluorescent pH sensors." *Chemical Communications* no. 19:2646–2648.

Wang, Ying, Shuo Wang, Jingjing Sun, Hengren Dai, Beijun Zhang, Weidong Xiang, Zixin Hu, Pan Li, Jinshui Yang, and Wen Zhang. 2021. "Nanobubbles promote nutrient utilization and plant growth in rice by upregulating nutrient uptake genes and stimulating growth hormone production." *Science of the Total Environment* no. 800:149627.

Yadav, Jayant, Poonam Jasrotia, Prem Lal Kashyap, Ajay Kumar Bhardwaj, Sudheer Kumar, Maha Singh, and Gyanendra Pratap Singh. 2021. "Nanopesticides: Current status and scope for their application in agriculture." *Plant Protection Science* no. 58 (1):1–17.

Yao, Chunyan, Yongzhi Qi, Yuhui Zhao, Yang Xiang, Qinghai Chen, and Weiling Fu. 2009. "Aptamer-based piezoelectric quartz crystal microbalance biosensor array for the quantification of IgE." *Biosensors and Bioelectronics* no. 24 (8):2499–2503.

Yu, Zhimin, Guohua Zhao, Meichuan Liu, Yanzhu Lei, and Mingfang Li. 2010. "Fabrication of a novel atrazine biosensor and its subpart-per-trillion levels sensitive performance." *Environmental Science & Technology* no. 44 (20):7878–7883.

Zahra, Zahra, Zunaira Habib, Hyeseung Hyun, and Hafiz Muhammad Aamir Shahzad. 2022. "Overview on recent developments in the design, application, and impacts of nanofertilizers in agriculture." *Sustainability* no. 14 (15):9397.

Zhao, Guangying, Shuangju Song, Chun Wang, Qiuhua Wu, and Zhi Wang. 2011. "Determination of triazine herbicides in environmental water samples by high-performance liquid chromatography using graphene-coated magnetic nanoparticles as adsorbent." *Analytica Chimica Acta* no. 708 (1–2):155–159.

Zhao, Hongyuan, Huina Ma, Xiaoguang Li, Binbin Liu, Runqiang Liu, and Sridhar Komarneni. 2021. "Nanocomposite of halloysite nanotubes/multi-walled carbon nanotubes for methyl parathion electrochemical sensor application." *Applied Clay Science* no. 200:105907.

Zhao, Lijuan, Jose R Peralta-Videa, Cyren M Rico, Jose A Hernandez-Viezcas, Youping Sun, Genhua Niu, Alia Servin, Jose E Nunez, Maria Duarte-Gardea, and Jorge L Gardea-Torresdey. 2014a. "CeO_2 and ZnO nanoparticles change the nutritional qualities of cucumber (*Cucumis sativus*)." *Journal of Agricultural and Food Chemistry* no. 62 (13):2752–2759.

Zhao, Yuan, Liqiang Liu, Dezhao Kong, Hua Kuang, Libing Wang, and Chuanlai Xu. 2014b. "Dual amplified electrochemical immunosensor for highly sensitive detection of *Pantoea stewartii* sbusp. *stewartii*." *ACS Applied Materials & Interfaces* no. 6 (23):21178–21183.

Zheng, Lei, Fashui Hong, Shipeng Lu, and Chao Liu. 2005. "Effect of nano-TiO_2 on strength of naturally aged seeds and growth of spinach." *Biological Trace Element Research* no. 104 (1):83–91.

Zulfiqar, Faisal, Míriam Navarro, Muhammad Ashraf, Nudrat Aisha Akram, and Sergi Munné-Bosch. 2019. "Nanofertilizer use for sustainable agriculture: Advantages and limitations." *Plant Science* no. 289:110270.

6 SNM in Food Preservation

6.1 SMART FOOD PACKAGING AND SECURITY

One of the most crucial steps in food safety is food packaging. All packaging materials are completely impermeable to natural substances, atmospheric gases, and water vapors. However, completely preventing gas migration and permeability is not preferred when packaging fresh fruits and vegetables, which go through cellular respiration. To prevent oxidation and de-carbonation, carbonated beverage packaging should eliminate the flow of oxygen and carbon dioxide (CO_2). The flow of CO_2, oxygen, and water vapors differs depending on the food matrix and the packaging materials used. As a result, by utilizing various nanocomposite materials, including polymers, these complexities in the food packaging can be addressed and overcome (Bajpai et al., 2018).

Food storage with preserved nutritional and organoleptic qualities is a crucial step in the food industry. Food packaging not only preserves the food from spoilage but also protects the sensitive bioactive compounds in foods from harsh physical and environmental conditions.

Several studies are being conducted using nanotechnology to improve the capabilities of food packaging. Various functional nanomaterials have been studied to enhance the properties of packaging materials such as barrier properties, thermal stability, strength, and durability, which help to extend the shelf life of perishable food products. In general, there are two ways to create nanotechnology-based packaging materials: either by adding nanomaterials to conventional food packaging materials like films and containers, or by creating multi-layer nanocomposite packaging materials via combining nanocoatings by immersion, spraying, or rubbing methods. The development of improved packaging materials is in three different forms; improved packaging with better mechanical and physical properties, active packaging with antibacterial, antioxidant, and UV absorption properties and smart packaging with monitored/controlled food conditions (Figure 6.1).

6.2 FOOD FUNCTIONALITY

Numerous applications of nanotechnology exist in the field of food technology. The use of nanomaterials typically occurs in one of two ways: first, they are used in packaging materials, sensors, and in the sanitization of food manufacturing facilities; second, they are directly incorporated into the food to create nutraceutical delivery systems and alter the rheological and optical properties of the products. Nanocarriers are used in the food processing industries as preservatives, bioactive compounds, color additives, flavor additives, anticaking agents (to prevent lump formation),

DOI: 10.1201/9781003366270-6

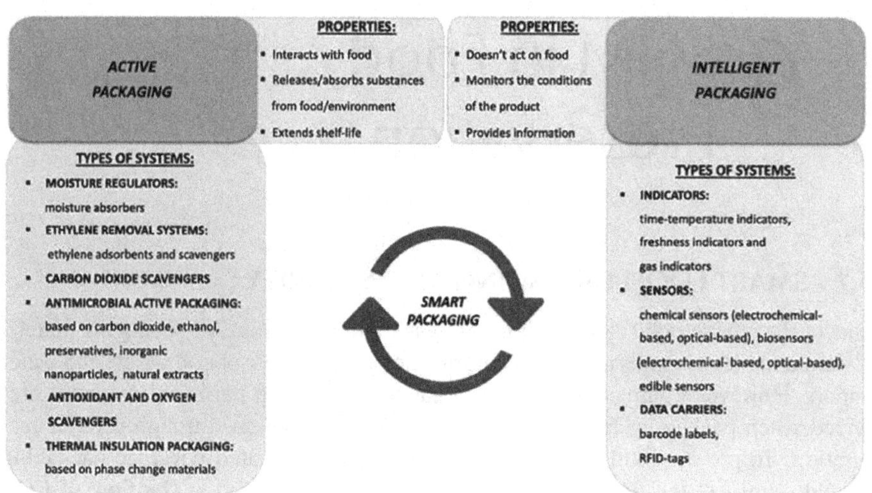

FIGURE 6.1 Active and intelligent packaging classification and their main properties. (Drago et al., 2020.)

gelling agents (to improve food texture), and color additives (Martínez-Ballesta et al., 2018). Many inorganic nanoparticle oxides such as TiO_2, SiO_2 (E551), copper oxide (CuO), iron oxide (Fe_2O_3), and zinc oxides (ZnO) are used as color additives, anticaking agent and nutritional supplements (Hossain et al., 2021).

Application of the nanoencapsulation protocol enhances the quality of food products. This method is frequently used to increase the flavor, preservation of food, and providing balance in cooking. Various techniques for nanoencapsulation of food bio actives are shown in Figure 6.2. When using plant oil instead of hydrogenated oil, trans fatty acids are reduced, which leads to the formation of safer nanofood production of nanocapsules for use in food to distribute nutrients for improved absorption. Nanoceramic in a pot-like form may be employed to modify absorption for decreasing the cooking time as well as the quantity of spent oil (Neme et al., 2021). Further, nanoencapsulation conceals tastes and odors, and controls how active ingredients interact with food, regulates the release of dynamic agents. Moreover, nanoencapsulation ensures the protection against moisture, chemical, heat, or biological interference during storage, processing, and applications (Singh et al., 2017). Metallic oxides like SiO_2 and TiO_2 have been used in food products as coloring or flow ingredients. In order to transport scents or aromas in food material, SiO_2 nanomaterials are applied in wasted food nanomaterials (Dekkers et al., 2011).

Nanoemulsions are a colloidal particulate system in the submicron size range acting as carriers (Jaiswal et al., 2015). They have some distinctive qualities that make them ideal in food industry formulas, including small size, increased surface area, and less sensitivity to physical and chemical changes (Salem and Ezzat, 2019). Due to the protection they provide against food splitting and breaking, nanoemulsions significantly reduce the amount of stabilizers needed. Many nanoemulsions appear optically clear and have numerous technical benefits for blending with beverages (Prasad et al., 2017). The final nanoemulsion products are very creamy like regular food products and do

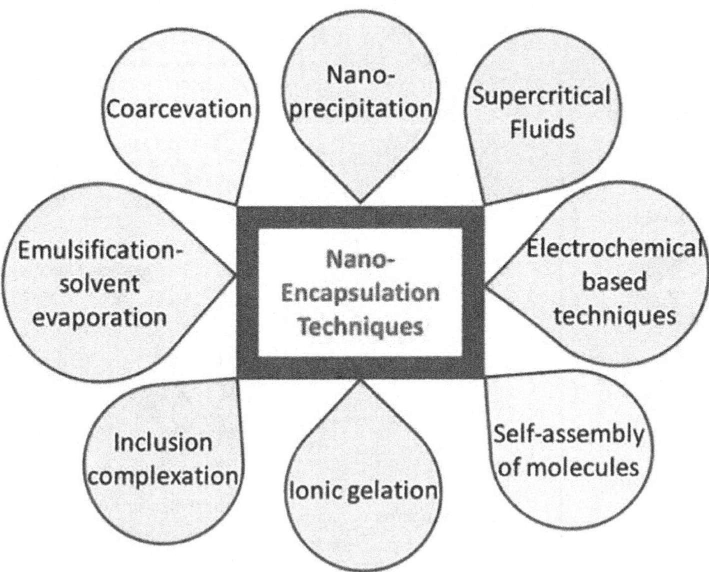

FIGURE 6.2 Techniques for nanoencapsulation of food bio actives. (Pateiro et al., 2021.)

not alter the mouthfeel or flavour and used in delivery systems and excipient foods as well (Figure 6.3). To modify food for salad dressing, the flavor in oils, sweeteners, improved beverages, and other food processing, nanoemulsions are used (Dasgupta et al., 2015). The encapsulation of lipophilic ingredients like vitamins, flavors, and nutraceuticals is one of the most significant uses of nanoemulsions in the food industry. Li et al., prepared nanoemulsions using various surfactants and co-surfactants. Five suitable surfactant mixtures were identified (Li et al., 2018). Ryu et al. studied the effect of ripening inhibitor types (corn oil, palm oil, coconut oil, and canola oil) on thyme oil emulsion stability, formation, and antimicrobial activity. A sufficient concentration of ripening inhibitor was determined to be around 40% of the oil phase, while continued antimicrobial activity during storage was evident (Ryu et al., 2018).

6.3 ENHANCE SHELF LIFE OF POSTHARVEST CROPS

Modern consumers prefer fresh foods due to their high nutritional value and demand high-quality fruits and vegetables that are rich in compounds that promote health (Zambrano-Zaragoza et al., 2021). Additionally, these foods must maintain their good physicochemical and sensory quality, be safe for consumption, and be free of contaminants and pathogenic microorganisms in order to promote more favorable health. The food and horticulture industries now face a challenge from this expanding demand to develop useful preservation techniques. Scientists and the food processing industries are now under pressure to evaluate various methods for improving the freshness, quality, shelf life, and food safety through the use of natural, edible, and biodegradable polymers (Tahir et al., 2019). These edible coatings and films create a semipermeable safety barrier around fruits, preventing the loss of quality

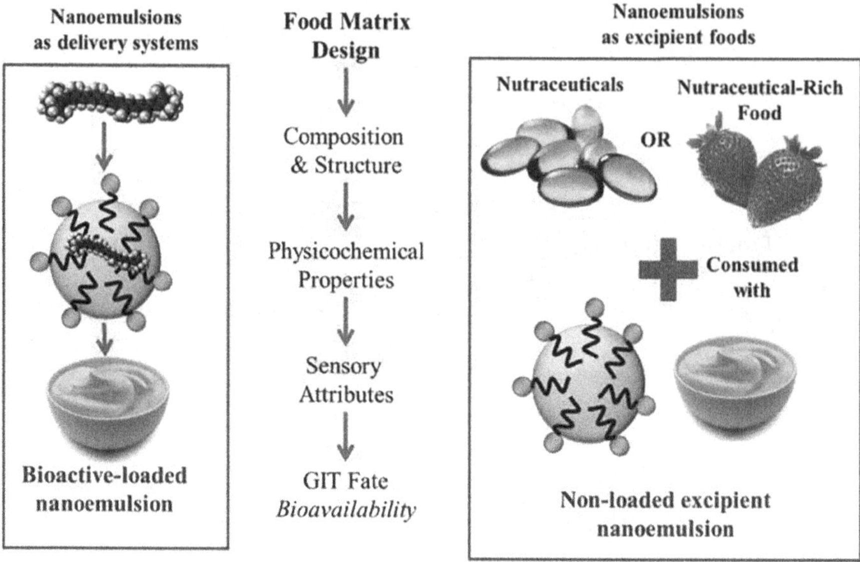

FIGURE 6.3 Schematic diagram of the difference between integrated and non-integrated excipient foods. For integrated excipient foods the bioactive component (pharmaceutical or nutraceutical) is dispersed within the excipient food matrix, but for non-integrated excipient foods the bioactive component is in another product that is co-ingested with the excipient food. (Salvia-Trujillo et al., 2016.)

characteristics. Polysaccharides, lipids, and proteins are the three different biological material types that can be used to create edible films and coatings. To create edible films and food coatings, a variety of biopolymers including alginate, carrageenan, chitosan, pectin, starch, and xanthan gum have been extensively used. Their film-forming properties allow the synthesis of membranes (thickness $> 30\,\mu m$) and coatings ($<30\,\mu m$), which are successfully used to preserve foodstuffs.

When antimicrobial agents are carried by edible coatings, microbial spoilage is replaced, extending the shelf life of the product and improving the functionality (Prakash et al., 2020) (Figure 6.4). According to Zaragoza et al., the edible coatings regulate both the rate of respiration and the growth of microbes during the preservation of fruits and vegetables (Zambrano-Zaragoza et al., 2021). Chitosan should be used in organic coatings to prevent food spoilage and the contaminations. It can absorb heavy metals with an extremely impressive capacity. By lowering transpiration and respiration rates, it is used as edible coatings to increase the shelf life of fruits. TiO_2 nanoparticles are highly effective at removing organic pollutants thanks to their excellent photocatalytic activity. The chemical grafting of antioxidant molecules (like chitosan) directly on the surface of TiO_2 nanoparticles has the best effect on the treatment of wastewater pollutants, and the integration of TiO_2 and chitosan can thus complement each other with their own advantages. Madhusha and co-workers have synthesized ascorbic acid-intercalated layered double hydroxides (LDHs) as an edible-coating for fresh food (Madhusha et al., 2020).

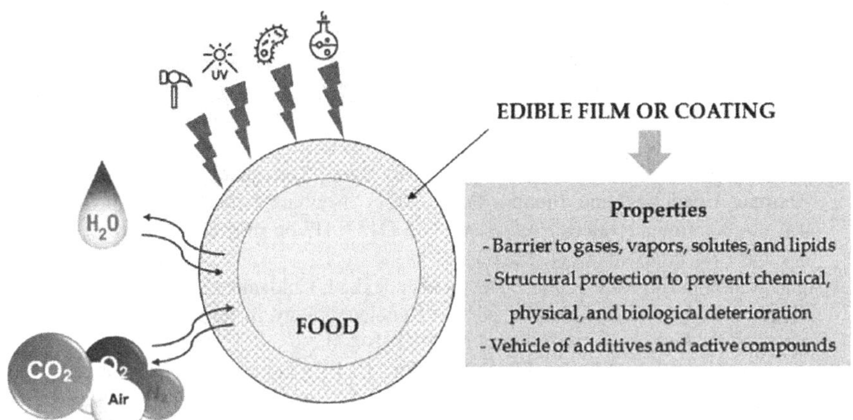

FIGURE 6.4 Main functions of edible films and coatings in food packaging applications. (Valdes et al., 2017.)

TABLE 6.1
Examples of Nanoedible Coatings in Food Products and Their Effect

Active Compound in Nanocoating	Applied Food Product	Effect	Reference
Carvacrol	Cabbage	Inhibition the growth of *P. Pastoris* and *E. coli*	Sow et al. (2017)
Carvacrol	Cucumber	Inhibition of the growth of *E. coli.*	Chen et al. (2021)
Lemongrass essential oil	Fresh-cut apple	Suppression of *E. coli* and sluggish development of psychrophilic microbes a	Gago et al. (2020)
Lemongrass oil	Grape berry	Inhibition the growth of *S. typhimurium*, moulds, yeast, and mesophiles	Oh et al. (2017)
Mandarin essential oil	Green beans	Reduction of the population of *L. innocua*	Severino et al. (2014)
Alfa tocopherol	Fresh-cut apples	Inhibitory effect on Pectin methyl esterase and polyphenol oxidase behavior	Zambrano-Zaragoza et al. (2014)

Reduced water loss, movement of fats and oils, gas diffusion, solute movement, losses of volatile flavors and aromas, improved structural properties (holding it together), and ease of incorporation of food additives, pigments, and flavoring agents are just a few of the significant benefits of using edible nano-coating. These coatings also prevent the transfer of oxygen and moisture, enhance the appearance of the food product, lessen the growth of mold and other fungi, and lessen the adhesion of food particles to the cooking surface (Mahela et al., 2020). More examples are shown in Table 6.1.

REFERENCES

Bajpai, Vivek K, Madhu Kamle, Shruti Shukla, Dipendra Kumar Mahato, Pranjal Chandra, Seung Kyu Hwang, Pradeep Kumar, Yun Suk Huh, and Young-Kyu Han. 2018. "Prospects of using nanotechnology for food preservation, safety, and security." *Journal of Food and Drug Analysis* no. 26 (4):1201–1214.

Chen, Chi-Hung, Hsin-Bai Yin, Zi Teng, Suyeun Byun, Yongguang Guan, Yaguang Luo, Abhinav Upadhyay, and Jitendra Patel. 2021. "Nanoemulsified carvacrol as a novel washing treatment reduces *Escherichia coli* O157: H7 on spinach and lettuce." *Journal of Food Protection* no. 84 (12):2163–2173.

Dasgupta, Nandita, Shivendu Ranjan, Deepa Mundekkad, Chidambaram Ramalingam, Rishi Shanker, and Ashutosh Kumar. 2015. "Nanotechnology in agro-food: From field to plate." *Food Research International* no. 69:381–400.

Dekkers, Susan, Petra Krystek, Ruud JB Peters, Daniëlle PK Lankveld, Bas GH Bokkers, Paula H van Hoeven-Arentzen, Hans Bouwmeester, and Agnes G Oomen. 2011. "Presence and risks of nanosilica in food products." *Nanotoxicology* no. 5 (3):393–405.

Drago, Emanuela, Roberta Campardelli, Margherita Pettinato, and Patrizia Perego. 2020. "Innovations in smart packaging concepts for food: An extensive review." *Foods* no. 9 (11):1628.

Gago, Custódia, Rui Antão, Cristino Dores, Adriana Guerreiro, Maria Graça Miguel, Maria Leonor Faleiro, Ana Cristina Figueiredo, and Maria Dulce Antunes. 2020. "The effect of nanocoatings enriched with essential oils on 'rocha'pear long storage." *Foods* no. 9 (2):240.

Hossain, Akbar, Milan Skalicky, Marian Brestic, Subhasis Mahari, Rout George Kerry, Sagar Maitra, Sukamal Sarkar, Saikat Saha, Preetha Bhadra, and Marek Popov. 2021. "Application of nanomaterials to ensure quality and nutritional safety of food." *Journal of Nanomaterials* no. 2021:9336082.

Jaiswal, Manjit, Rupesh Dudhe, and PK Sharma. 2015. "Nanoemulsion: An advanced mode of drug delivery system." *3 Biotech* no. 5 (2):123–127.

Li, Ze-hua, Ming Cai, Yuan-shuai Liu, and Pei-long Sun. 2018. "Development of finger citron (*Citrus medica* L. var. *sarcodactylis*) essential oil loaded nanoemulsion and its antimicrobial activity." *Food Control* no. 94:317–323.

Madhusha, Chamalki, Imalka Munaweera, Veranja Karunaratne, and Nilwala Kottegoda. 2020. "Facile mechanochemical approach to synthesizing edible food preservation coatings based on alginate/ascorbic acid-layered double hydroxide bio-nanohybrids." *Journal of Agricultural and Food Chemistry* no. 68 (33):8962–8975.

Mahela, U, DK Rana, U Joshi, and YS Tariyal. 2020. "Nano edible coatings and their applications in food preservation." *Journal of Postharvest Technology* no. 8 (4):52–63.

Martínez-Ballesta, MCarment, Ángel Gil-Izquierdo, Cristina García-Viguera, and Raúl Domínguez-Perles. 2018. "Nanoparticles and controlled delivery for bioactive compounds: Outlining challenges for new "smart-foods" for health." *Foods* no. 7 (5):72.

Neme, Kumera, Ayman Nafady, Siraj Uddin, and Yetenayet B Tola. 2021. "Application of nanotechnology in agriculture, postharvest loss reduction and food processing: Food security implication and challenges." *Heliyon* no. 7 (12):e08539.

Oh, Yoon Ah, Yeong Ji Oh, Ah Young Song, Jin Sung Won, Kyung Bin Song, and Sea C Min. 2017. "Comparison of effectiveness of edible coatings using emulsions containing lemongrass oil of different size droplets on grape berry safety and preservation." *LWT* no. 75:742–750.

Pateiro, Mirian, Belén Gómez, Paulo ES Munekata, Francisco J Barba, Predrag Putnik, Danijela Bursać Kovačević, and José M Lorenzo. 2021. "Nanoencapsulation of promising bioactive compounds to improve their absorption, stability, functionality and the appearance of the final food products." *Molecules* no. 26 (6):1547.

Prakash, Anand, Revathy Baskaran, and Vellingeri Vadivel. 2020. "Citral nanoemulsion incorporated edible coating to extend the shelf life of fresh cut pineapples." *LWT* no. 118:108851.

Prasad, Ram, Atanu Bhattacharyya, and Quang D Nguyen. 2017. "Nanotechnology in sustainable agriculture: Recent developments, challenges, and perspectives." *Frontiers in Microbiology* no. 8:1014.

Ryu, Victor, David J McClements, Maria G Corradini, and Lynne McLandsborough. 2018. "Effect of ripening inhibitor type on formation, stability, and antimicrobial activity of thyme oil nanoemulsion." *Food Chemistry* no. 245:104–111.

Salem, Mohamed A, and Shahira M Ezzat. 2019. "Nanoemulsions in food industry." In Jafar M. Milani (Ed.), *Some new aspects of colloidal systems in foods*, 32–51, IntechOpen, London.

Salvia-Trujillo, Laura, Olga Martín-Belloso, and David Julian McClements. 2016. "Excipient nanoemulsions for improving oral bioavailability of bioactives." *Nanomaterials* no. 6 (1):17.

Severino, Renato, Khanh Dang Vu, Francesco Donsì, Stephane Salmieri, Giovanna Ferrari, and Monique Lacroix. 2014. "Antibacterial and physical effects of modified chitosan based-coating containing nanoemulsion of mandarin essential oil and three non-thermal treatments against *Listeria innocua* in green beans." *International Journal of Food Microbiology* no. 191:82–88.

Singh, Trepti, Shruti Shukla, Pradeep Kumar, Verinder Wahla, Vivek K Bajpai, and Irfan A Rather. 2017. "Application of nanotechnology in food science: Perception and overview." *Frontiers in Microbiology* no. 8:1501.

Sow, Li Cheng, Felisa Tirtawinata, Hongshun Yang, Qingsong Shao, and Shifei Wang. 2017. "Carvacrol nanoemulsion combined with acid electrolysed water to inactivate bacteria, yeast in vitro and native microflora on shredded cabbages." *Food Control* no. 76:88–95.

Tahir, Haroon Elrasheid, Zou Xiaobo, Gustav Komla Mahunu, Muhammad Arslan, Mandour Abdalhai, and Li Zhihua. 2019. "Recent developments in gum edible coating applications for fruits and vegetables preservation: A review." *Carbohydrate Polymers* no. 224:115141.

Valdes, Arantzazu, Marina Ramos, Ana Beltrán, Alfonso Jiménez, and María Carmen Garrigós. 2017. "State of the art of antimicrobial edible coatings for food packaging applications." *Coatings* no. 7 (4):56.

Zambrano-Zaragoza, Maria L, Elsa Gutiérrez-Cortez, Alicia Del Real, Ricardo M González-Reza, Moises J Galindo-Pérez, and David Quintanar-Guerrero. 2014. "Fresh-cut Red Delicious apples coating using tocopherol/mucilage nanoemulsion: Effect of coating on polyphenol oxidase and pectin methylesterase activities." *Food Research International* no. 62:974–983.

Zambrano-Zaragoza, María L, David Quintanar-Guerrero, Ricardo M González-Reza, María A Cornejo-Villegas, Gerardo Leyva-Gómez, and Zaida Urbán-Morlán. 2021. "Effects of UV-C and edible nano-coating as a combined strategy to preserve fresh-cut cucumber." *Polymers* no. 13 (21):3705.

7 SNM in Apparel Industry

7.1 CURRENT RESEARCH TRENDS FOR SMART TEXTILE

Animal skins or plant leaves were employed by our ancestors to create clothing, mostly as a form of defense against harsh environmental conditions. Natural fibers like flax, silk, and cotton were gradually fashioned into more comfy clothes. More recently, a wide variety of synthetic polymers, including nylon, polyester, and acrylic gradually came into usage. The last 10 years have seen new obstacles for the textile industry, including advancements in electrical miniaturization and novel intelligent materials (Ferri et al., 2019). In terms of hydrophobicity (wearer comfort), UV resistance, antibacterial, antistatic, anti-wrinkle, stain-free, or shrink resistance, modern fabrics outperform "old" textiles. These are "passive features," however, and researchers are eager to employ nanotechnology to infuse ingenious and imaginative applications by using novel fabrication and surface finishing techniques. Their main objective is to create new, extremely useful applications while preserving the comfort, flexibility, and light weight of the cloth (Kirstein, 2013).

7.2 ANTIMICROBIAL TEXTILES

Antibacterial protection offered by textiles is quite intriguing and advantageous to human health. TiO_2, chitosan, N-halamine, $AgCu_2O$, and metal/hemp fibers, among other antimicrobial compounds, have all been introduced into fabrics for their antibacterial properties (Shah et al., 2022). The active nanomaterials can be chemically or physically integrated into the fabrics to create antibacterial textiles. In order to create antimicrobial nanomaterials, Muoz-Bonilla and Fernández-Garca employed a variety of techniques, including electrospinning, nano-precipitation, and self-assembly (Muñoz-Bonilla and Fernández-García, 2015.)

Silver (Ag) is one of the most primitive antibacterial nanoparticles used on textile surfaces. Without altering its mechanical qualities, it serves as a doping antibacterial agent and demonstrates exceptional antimicrobial efficacy (Vasantharaj et al., 2019). Because of their small size and large surface area, Ag nanoparticles interact with bacterial proteins to stop the growth of their cells. Ag nanoparticles additionally obstruct the system responsible for transporting electrons and substrates. The Ag^+ ions formed when moisture reacts with an organism quickly and diffuses its cell membrane and cell wall and enter the cytoplasm. The S-containing proteins on the cell membrane interact with the Ag^+ ions, changing the shape of the cell wall. As a result, osmotic activity damages the cell membrane and causes it to leak cytoplasm. Additionally, the Ag^+ ions work with the proteins that contain phosphate to condense DNA, which ultimately results in cell death (Wilkinson et al., 2011). Ag nanoparticles' ability to inhibit microbial growth depends on their size, surface area, concentration, and generation of Ag^+ ions. Patil and colleagues looked at a quick,

DOI: 10.1201/9781003366270-7

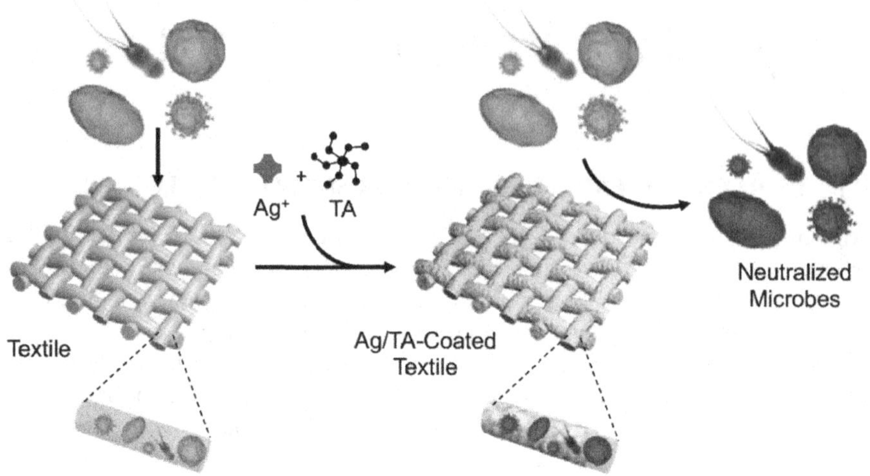

FIGURE 7.1 Schematic illustration of the formation of colorless Ag/TA coating on textiles and their interactions with microbes. While pathogenic and odor-causing microbes can survive on uncoated textiles, Ag/TA-coated textiles rapidly neutralize a wide range of microbes. (Richardson et al., 2022.)

one-step sono-chemical synthesis and deposition process to make cotton nanoparticles with Ag coatings (Patil et al., 2019). Richardson et al. have also demonstrated the same phenomenon (Figure 7.1). They showed that the Ag NPs have the strongest antibacterial activity and are stable, mono-dispersed, uniformly deposited on cotton fibers. Additionally, the antibacterial and self-cleaning properties of Ag-doped SiO_2 nanoparticles with core corona shape on cotton fabrics were investigated. By coating these corona-structured nanoparticles with antibacterial substances like quaternary ammonium salts, the bacteria can be made inactive. Reactive oxygen species (ROS) such as superoxides, hydroxyl radicals, or a positive hole can be produced by TiO_2. These ROS may interact with the bacterial cell wall and membrane, ultimately resulting in cell death. Antibacterial fabrics have made use of this TiO_2 nanoparticle feature. The ROS can also break down organic matter or greasy filth, giving textiles the ability to self-clean. If TiO_2 is doped with additional active species like Ag, Au, SiO_2, ZnO, etc., this self-cleaning property can be strengthened even more.

Riaz and colleagues looked into how TiO_2 may be used in the textile sector along with 3-(trimethoxysilyl) propyl N,N,N-dimethyloctadecylammonium chloride and 3-glycidoxypropyltrimethoxysilane (Riaz et al., 2019). The researchers have come to the conclusion that treated cotton exhibited long-lasting super-hydrophobicity, self-cleaning, and antibacterial activity. ZnO nanoparticles exhibit antibacterial and self-cleaning capabilities for textiles loaded with gram-negative bacteria, and they behave similarly to TiO_2 nanoparticles. The fabrication of ZnO nanoparticles and their integration into cotton fabrics were the results of Patil and colleagues' investigation into sonochemical synthesis procedures (Patil et al., 2019). The cotton fabrics that had ZnO nanoparticles applied to them demonstrated flexural stiffness, tensile strength, water contact angle, and air permeability. They demonstrated considerable antibacterial activities and outstanding nanoparticle deposition characteristics on cotton fabric yarns. For antibacterial activity and UV protection, Fouda and colleagues mixed bio-active macromolecules released by bio-synthesized ZnO and fungal nanoparticles (Fouda et al., 2018). They used an isolated fungus called *Aspergillus terreus* to extract proteins that are attracted to the caps of ZnO nanoparticles. They discovered that coated textiles with biosynthesized ZnO nanoparticles could prevent the growth of pathogenic bacteria in comparison to untreated materials. Green synthesis was used by Karthik and colleagues to synthesize ZnO nanoparticles that had a noticeable antibacterial effect (Karthik et al., 2017). Salat and colleagues have worked on the coating of cotton medical textiles with gallic acid and antibacterial ZnO nanoparticles (Salat et al., 2018). They have shown that gallic acid enables the coated materials to safely come into touch with the phenolic network, an antibacterial agent, and human skin. Utilizing electrospinning technique, Yu and colleagues created core-spun nanofiber yarns with exceptional antibacterial qualities (Yu et al., 2021). Nearly 100% of the yarn structure has antibacterial properties. With the aid of the ultrasonication approach and green synthesis, Hiremath and colleagues synthesized magnetite nanoparticles that effectively provide protection against microbes (Hiremath et al., 2018).

Since the COVID-19 outbreak, face masks made of nanomaterial have received a lot of attention. Researchers from several fields created antiviral face masks and personal protective equipment (PPE) kits that could filter out a variety of diseases, including SARS-CoV-2. Talebian and colleagues (2020) offered two strategies to regulate COVID-19 involving biosensors on mask or PPE fabrics and disinfectants based on nanomaterials (Talebian et al., 2020). Researchers suggest that metallic nanoparticles such as Ag, Cu, TiO_2, etc. can be alternatives to the traditional disinfectants such as chlorides, quaternary amines, peroxides, and alcohols. They also propose that highly efficient biosensors can be integrated on face masks or PPE kits so that early detection of SARS-CoV-2 or other viruses can be realized. These face masks are proposed to prevent spread of virus via sneezing and coughing.

7.3 HYDROPHOBICITY AND OLEOPHOBICITY IN TEXTILES

The true creator of intelligent and smart materials is nature. It has frequently motivated scientists to emulate biological events. In the case of the hydrophobicity phenomenon, the same scenario can be observed. For instance, the preening oil coating on the feathers of the ducks allows them to survive in the water (Liu et al., 2008). The researchers use chitosan coatings over cotton and polyester textiles to imitate this natural process. The chitosan coating solution is prepared using a precipitation technique and then further treated with a silicone compound to produce a surface energy that is lower. Similarly, to create superhydrophobic surfaces, the researchers have used virgin and surface-modified carbon nanotubes (CNTs) on cotton fibers. They have done this to emulate the roughness of lotus leaves. Over 150° of significant contact angle was attained. Ramaratnam and colleagues carried out another similar study in which hydrophobic nanocoatings (20 nm) were developed in order to produce hydrophobic fabrics (Ramaratnam et al., 2007). Nanowhiskers mounted on hydrocarbons can also be used to create fibers that repel water. These materials are approximately one-third the size of traditional cotton fibers. To create a peach fuzz look, these nanowhiskers can be added to textile fibers. The distance between each individual nanowhisker is more than the molecular size of H_2O but smaller than the size of a water droplet. As a result, there may be noticeable surface tension, which prevents water from spreading on its surface. However, the permeability of nanowhiskers allows for the maintenance of breathability (Zahid et al., 2019). Consequently, nanoparticulate films on the textiles can be used to create water repellent coatings. For use with textile polymers, fluorinated combinations are frequently utilized. Superhydrophobicity can be achieved by properly treating fibers to adjust their texture, without sacrificing the materials' comfort, softness, or toughness. Das and co-workers have revealed formation of fluorinated silyl functionalized superhydrophobic zirconia coating on cotton fabric (Das and De, 2015) (Figure 7.2).

To achieve hydrophobicity or oleophobicity, contact angle tuning is essential. When SiO_2 nanoparticles (143–378 nm) were combined with a water-repelling substance, a substantial contact angle of more than 130° was reported. SiO_2 nanoparticles can also be used along with perfluorooctylated quaternary ammonium silane (PQAS) as the coupling agent (Yu et al., 2007). A decent contact angle of 145° was reported which led to excellent hydrophobicity, owing to the diminishing of surface energy by PQAS. The oleophobicity was also enhanced; exhibiting a contact angle of 131° when a droplet of diiodomethane (CH_2I_2) was used on the surfaces of fabrics. Amphiphilic janus type micro/nanoparticles were also deposited on the fabric surfaces to achieve hydrophobicity (Synytska et al., 2011). The microparticles help in crosslinking between the fibers, while the nanoparticles stick to the surface of fiber.

7.4 ULTRAVIOLET RESISTANT TEXTILES

To improve UV shielding, fabrics are treated with UV-blocking nanomaterials (UVB and UVA radiations) to create UV protection materials. The ultraviolet protection factor (UPF), which is dependent on the fabric's composition, is a measurement tool for the effectiveness of UV protection (Yang et al., 2004). UV radiation and how it works in a textile structure is illustrated in Figure 7.3.

TiO_2 and ZnO are two examples of common nanomaterials that can scatter or absorb UV rays. These substances can remain stable even at greater temperatures and are both stable and non-toxic. The scattering of UV light by nanoparticles depends on the wavelength of the radiation and the size of the particles. TiO_2 nanoparticles have been utilized as UV blockers on cotton. Even after 50 washings, the TiO_2 finishing's durability was confirmed to be good (Daoud and Xin, 2004). Additionally, ZnO nanorods have been employed to create an effective UV scattering layer on cotton fabrics. Additionally, ZnO nanoparticles have been used as a UV-absorbing layer on cotton and polyester fabrics. The anti-UV qualities of polyaniline/titanium dioxide (PANI/TiO_2) and PANI cotton fabrics were confirmed by Yu and colleagues. MnO_2-$FeTiO_3$ nanoparticles combined with thermoplastic polyurethane cotton fabrics are reported to aid in blocking of UV radiation (Dhineshbabu and Bose, 2019). These findings show that, in comparison to the uncoated materials, nanocoated materials on textiles have great UV-blocking capacity, making them more intelligent and

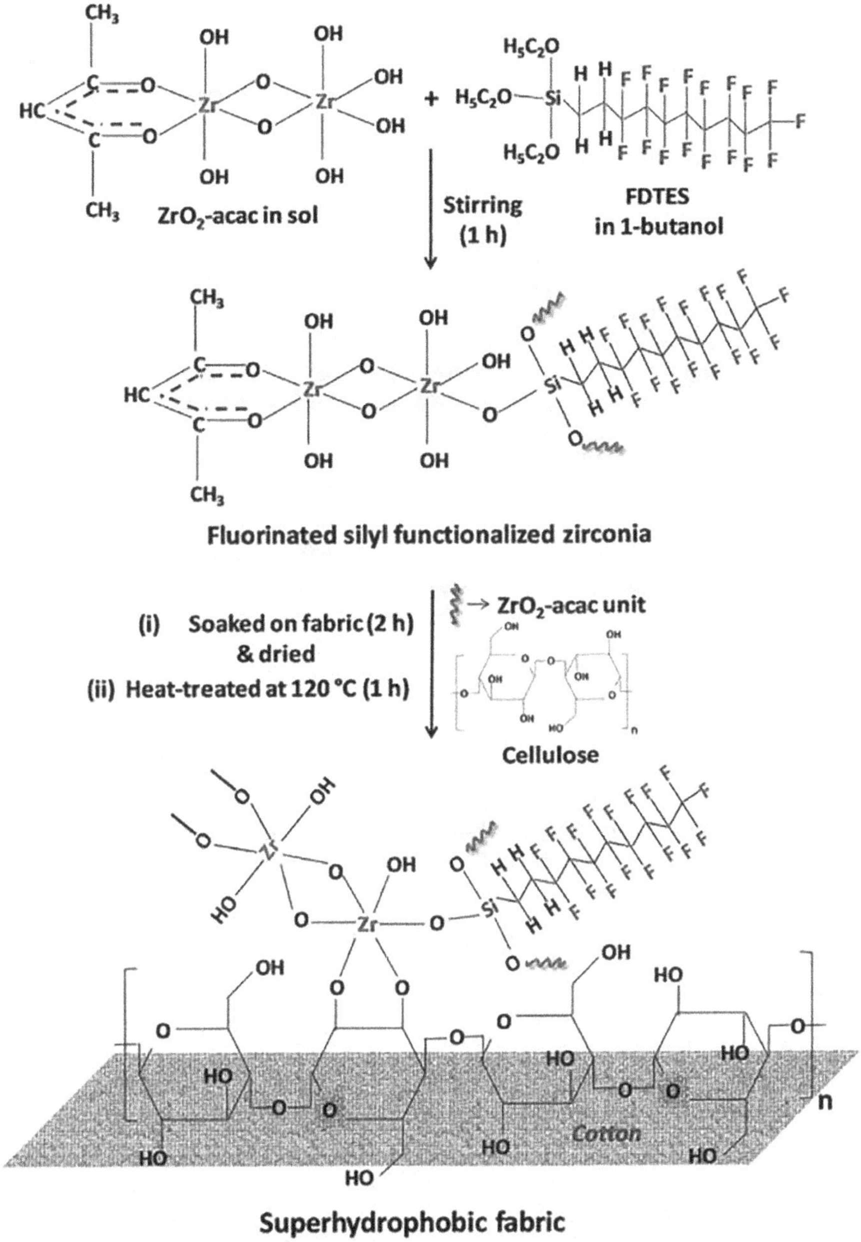

FIGURE 7.2 Schematic representation of fluorinated silyl functionalized superhydrophobic zirconia coated fabric. The reaction principle shows the formation of fluorinated silyl functionalized superhydrophobic zirconia coating on cotton fabric. (Das and De, 2015.)

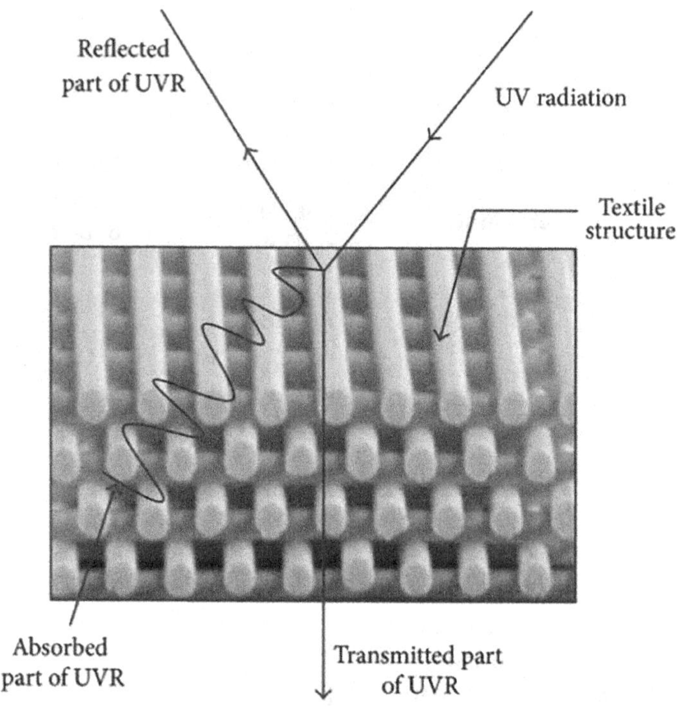

FIGURE 7.3 UV radiation and textile structure. (Singh and Singh, 2013.)

long-lasting fabrics. The UV-absorbing phenomenon has several uses in textiles because it can help people to shield from hazardous UV exposure.

7.5 ANTISTATIC PROPERTIES IN TEXTILES

Nylon and polyester, being hydrophobic, exhibit larger static charges. In contrast, the higher moisture content of cellulosic fibers lowers their static charges. To obtain antistatic capabilities in synthetic fibers, a variety of nanomaterials have been used, including ZnO whiskers, TiO_2 nanoparticles, Sb-doped SnO_2 nanoparticles, etc. (Zhou et al., 2003). Due to their conductive nature, these nanoparticles disperse the static charge on the fabrics. Due to their ability to interact with the hydroxyl groups on the surface of the air to absorb moisture, some nanosols based on silanes have also been utilized as antistatic agents. Commercially, a conductive nanoparticle-coated poly(tetrafluoroethylene) (PTFE) antistatic membrane has been produced (Shishoo, 2002). To obtain antistatic qualities, several researchers created sol-gel coatings on the surface of the fibers. After being modified with hydrophilic compounds or alkoxysilanes that include amino groups, several hydrophobic chemical species, such as alkoxysilanes, are also used. Textiles with a sol-gel coating have antistatic qualities because the coatings are hydrophobic on the surface but contain moisture underneath. In polyester fabric, Ag nanoparticles with a fluorine hydrophobic coating can produce antistatic characteristics. It has also been noted that ZnO nanoparticle coatings exhibit antistatic properties while Ag nanoparticles could decrease the static voltage of polyester fiber by 60.4%. When Au and ZnO nanoparticles were mixed, the static voltage dropped by 77.7%. Sb-doped SnO_2 for antistatic characteristics in polyacrylonitrile (PAN) fibers was also studied (Wang et al., 2004). When these nanoparticles diffused into the fibers, conductive pathways were created, which finally resulted in antistatic properties.

7.6 ELECTRICALLY CONDUCTIVE TEXTILES

The conductive properties of the nanomaterial are the key factors driving the introduction of actuators and sensors in the textile sector. Conducting polymers are used in a wide range of applications in the textile sector. When exposed to an external stimulus, the textile surface of these materials responds electrically due to the tuning of their resistivity. By adding different nanomaterials to the polymer's matrix, these polymers can be altered to have a desired property. For instance, conducting polymers with higher mechanical strength, optical, and conducting properties include nanostructured PANI, polypyrrole (PPy), and polythiophene (PT). For instance, conducting polymers with higher mechanical strength, optical, and conducting properties include nanostructured PANI, PPy, and PT (Anbarasan et al., 1999).

To change the surface structure of the fibers and enable diverse smart functions, several conductive nanomaterials have been developed. Conductive polymers applied to fiber surfaces increase the conductivity of such fibers by an order of magnitude. For instance, spinning has been used to combine SiO_2 nanoparticles with polyimidoamide fibers. The introduction of nanoparticles into PAN fibers resulted in the development of electrically conductive channels in the fibers. This resulted in more advanced mechanical and antistatic properties. The fibers have been coated with SiO_2 nanoparticles, diamine (diaminodiphenylmethane), and montmorillonite to increase their tenacity and thermal resistance (Yetisen et al., 2016). By using chemical oxidative deposition, conductive polymers like PANI, PPy, and PT can be employed to impart increased tensile strength and thermal stability in the synthetic fiber. These composite fibers have numerous uses in microwave attenuation, static electrical charge reduction, and electromagnetic shielding. Cotton has been coated with a variety of conductive matrices to add electrical conductivity. For conducting textiles, Shim and colleagues created a polyelectrolyte-based covering combined with multiwalled CNTs (Shim et al., 2008). In their study, Mattana and colleagues used a blend of several metal nanoparticles that were applied in a conformal manner around the heterogeneous contour of cotton fibers (Mattana et al., 2011). In-situ polymerization can be used to reduce mechanical deformations in cotton-based transistors because it creates flexible bridges between the nanoparticles. In order to incorporate the conductive properties, graphene has also been added to textile fibers. For instance, a fabric has been created by weaving together two sets of graphene micro-ribbons. The textiles' durability as-prepared was good. By adjusting the packing ribbon's density, the conductivity of this fabric was adjusted and optimized. Cu meshes were used as the substrate for the atmospheric chemical vapor deposition (CVD), which produced graphene fibers with wires that were around 60m in diameter. This was similar to how standard dip and dry techniques can immobilize graphene on a fabric. With this technique, graphene oxide was converted into graphene, and several layers were created, increasing the fabric's conductivity by up to three times. The right reducing agent and its concentration can be used to adjust the surface conductivity. In this instance, the graphene-coated cotton fabric had an electrical resistance of 103–106 KΩ cm^{-1}. In order to achieve a dispersion of short-sized CNT in water-based paste for the creation of electro-conductive fabrics, Trovato and colleagues created a flexible and novel approach (Trovato et al., 2020). They produced wearable conductive materials and demonstrated that nanotubes were evenly distributed on coatings. This implies that a variety of conducting 2D and 3D nanomaterials, which can be formed into wires, films, or coatings on textile fibers, may find several uses in on-body electronics.

7.7 ENERGY STORAGE BY TEXTILES

Supercapacitors have been used in textile technology for energy storage purposes. Researchers are trying to find a way to incorporate the supercapacitor electrodes into textiles without compromising its flexibility or wearability (Jost et al., 2011). Utilizing poly(methyl methacrylate) (PMMA) and polyethylene glycol, activated carbon has been used to alter cotton and polyester fabrics (PEG). In order to organize the supercapacitor cells in a traditional symmetrical two-electrode layout, screen printing was used on polyester microfibers. The cotton/polyester electrodes with activated carbon coating displayed a gravimetric and areal capacitance of 85 Fg^{-1} at 0.25 Ag^{-1}. Recently, an emulsion-electrospinning approach was used by Zhou and colleagues (2021) to generate in situ cross-linked polyvinyl alcohol/phase (PVA/PCM) nanofiber materials. They discovered that PVA/PCM nanofibers, when compared to conventional PVA/PCM nanofibers, have substantial applications in heat storage and temperature management due to their superior durability, energy storage, increased water resistance, thermal stability, and tensile strength. Lai and colleagues investigated a novel approach to easily dip-coat the wire-shaped solid-state supercapacitors using a soft aerogel (Lai et al., 2019). They electrospun hydrophilic PAN nanofibers with glycerol on titanium metal wire to create the sacrificial aerogel with a significant amount of void space. They demonstrated that polystyrene-sulfonate (PSS) etching may result in a mesoporous shape and that the capillary effect in the natural drying process can impede the dissolution of the template in the solvent. The researchers concluded that Ti/poly (3,4-ethylenedioxythiophene) (PEDOT) is an extremely potent source for wearable electronics. Using CNT/PANI composite fiber, Pan and colleagues created a flexible supercapacitor mounted textile. These textile supercapacitors were capable of photoelectric conversion and multilayer stacking to store energy (Pan et al., 2015). After being created by CVD, the CNTs were woven into the fibers by first creating a thicker coating through stacking. PANI was electrodeposited onto the textile fiber after it had been produced to create an electrode. The electrode was coated with a gel electrolyte to make a supercapacitor. Even after 200 cycles of bending, the material maintained a 96% capacitance of 272 F g^{-1}. Zhang and colleagues created supercapacitors, which kept a metal wire at the center of the CNT yarn, in an effort to improve the performance of the fabric (Zhang et al., 2014).

Using CNT/PANI composite fiber, Pan and colleagues created a flexible supercapacitor mounted textile (Pan et al., 2015).

Smart, energy-efficient triboelectric nanogenerators have also been built on wearable textiles. Triboelectric nanogenerators have also been created using the polydimethylsiloxane (PDMS) nanopatterns formed over ZnO nanorod arrays. These devices produce 120 V at 65 A, although their four-layered construction is capable of producing 170 V at 120 A. Even after 120,000 cycles, there was barely any deviation, proving their stability (Seung et al., 2015).

7.8 PHOTONICS IN TEXTILES

In order to regulate or research the look or other aspects of such textiles, photonic textiles integrate light emitting or light processing devices into a mechanically flexible matrix of a woven material, as implied by their name. Through the incorporation of specialized optical fibers during the weaving process of textile production, photonic textiles are practically implemented (Tao, 2005). This strategy makes sense since optical fibers, which are long threads with a sub-millimeter diameter, are mechanically and geometrically similar to conventional textile fibers and can be processed in the same manner. Numerous uses of photonic textiles have been investigated, including flexible and wearable screens, large-area illumination, and clothing with distinctive aesthetic appeal. Other uses include wearable sensing and large-area structural health monitoring. For in-service structural health monitoring and stress-strain monitoring of industrial textiles and composites, optical fibers embedded in woven composites have been employed. Wearable clothing that incorporates optical fiber-based sensor elements enables real-time monitoring of environmental and physical conditions, which is important for military and other dangerous civilian vocations. Optical fibers with chemically or biologically activated claddings for biochemical detection, Bragg gratings and long-period gratings for temperature and strain measurements, as well as micro bending-based sensing elements for pressure detection, are some examples of such sensor elements. Optical fiber sensors (OFSs) have several advantages over other sensor types, including durability to corrosion and wear, flexibility and light weight, immunity to interference from electricity and magnetism (E&M), and simplicity of textile integration.

Due to its potential use in dynamic signs and wearable advertising, flexible displays based on emissive fiber textiles have recently attracted a lot of attention. Such displays have been reported to be natural "attention grabbers," however, and may not be appropriate for applications that do not necessitate constant user awareness. The so-called ambient displays, which are based on non-emissive or perhaps weakly emissive elements, offer an alternative to such displays. In such displays, color changing is often accomplished in the light reflection mode using the chromatic inks' varying the spectrum absorption. Such inks have the ability to alter color or transparency by thermal or electrical activation. An ambient display typically blends in with the surroundings, and the user typically isn't even aware of its presence (Kirstein, 2013). It is suggested that comfort, esthetics, and information streaming are the easiest to integrate with these ambient displays.

Commercial total internal reflection (TIR) optical fibers created for the telecommunications sector are used in conventional fiber optic textiles. These fibers are designed for minimal loss of information transmission and are constructed of transparent materials like glass or polymers. Industrial textiles and composites with TIR optical fibers incorporated have been used for in-service structural health monitoring and stress-strain monitoring. Wearable clothing that incorporates optical fiber-based sensor elements enables real-time monitoring of environmental and physical conditions, which is crucial for many hazardous civil jobs, such as the security and military sectors. For biochemical detection, optical fibers with biologically or chemically activated claddings, Bragg gratings, long-period gratings, pressure-based micro bending sensors, and long-period gratings are a few examples of such sensor elements. The OFSs have advantages over other sensor types, such as resilience to corrosion and fatigue, flexibility and light weight, immunity to E&M interference, and simplicity of textile integration. Many technical issues with light extraction from optical fibers are solved when (photonic bandgap) PBG fibers are used instead of TIR fibers, and additional capabilities are also made possible.

7.9 SELF-CLEANING TEXTILES

The self-cleaning concept has drawn a lot of attention due to its unique qualities and large range of potential applications in a variety of industries (El-Khatib, 2012). In addition to the rising demand for hygienic, self-disinfecting, and contamination-free surfaces, interest in self-cleaning protective materials and surfaces have grown quickly over the years due to its high potential as a commercial good that can satisfy the expectations of the worldwide market. There are several materials that make use of the self-cleaning technology, including roof tiles, car mirrors, and solar panels as well as interior uses including fabrics, furniture, and window glass. The self-cleaning notion was inspired by a natural occurrence that may be seen on fish scales, butterfly wings, lotus leaves, rice plant leaves, and more. A species of plant known as a lotus leaf may thrive in mud without the dirt affecting thc plant's purity. The lotus leaves have exceptional hydrophobic surfaces due to their waxy coatings and the presence of tiny structures. Barthlott conducted research on the plant leaves' surface self-cleaning capacity in 1997 (Barthlott and Neinhuis, 1997). Despite being known since the 1970s, the 'Lotus Effect' was coined by Barthlott and his team in the 1990s as the cause of the self-cleaning qualities. It is based on the unique characteristics of extremely hydrophobic micro- and nanostructured surfaces, which are always completely cleaned by rainfall: the double structured surface greatly reduces the contact area between water and dirt particles. This, along with hydrophobic chemistry produces extraordinarily high contact angles that allow water droplets to roll off at the smallest inclination, picking up and eliminating all adhering particles and leaving behind a clean and dry surface in the process (Yuranova et al., 2006).

When exposed to either natural or artificial rain, water-repellent leaves always completely remove all kinds of particles typically as long as the surface waxes are not damaged. The dirt that accumulates on the waxy surface of the leaves is typically greater than the microstructure of the leaf's surface. Consequently, they are deposited on the tips, which reduce the interfacial area between the two surfaces. When a water droplet rolls over a particle, the surface area that is exposed to the air is decreased and energy is gained by adsorption. Only until a larger force dislodges the particle from the water droplet's surface will the adhesion between the particle and droplets be broken. Adhesion is reduced as a result of the extremely limited interfacial area between the particle and the rough surface. As a result, the water droplet "captures" the particle and removes it from the surface. The "Lotus effect" is the name of this phenomenon. Self-cleaning textiles are those that have a surface that can be cleaned by them without the need for laundering. Hence, the economic significance of the self-cleaning textiles can be stressed by the ease of maintenance and environmental protection because of reduced cleaning efforts and improved aging behavior by extended surface purity effect. Additionally, military personnel must endure conditions so harsh that they are unable to wash their clothes.

The two main mechanisms by which materials self-clean are hydrophobicity and hydrophilicity. With the help of water, both coating types may clean themselves by rolling water droplets for hydrophobic coatings and sheeting water for hydrophilic coatings, which removes dirt. However, hydrophilic coatings also possess the ability to chemically degrade the adsorbed dirt in sunlight with the aid of a photocatalyst, also referred to as a hydrophilic photocatalytic coating.

TiO_2 is a photocatalyst that has been shown to be an efficient catalyst in the photodegradation of colorants and other organic contaminants. It is widely utilized due to its several benefits, including non-toxicity, availability, cost effectiveness, chemical stability, and favorable physical and chemical qualities. Pisitsak and colleagues evaluated the self-cleaning properties of cotton fabrics treated with nano-TiO_2 and nano-TiO_2 combined with fumed silica. Self-cleaning was more effective in samples coated with greater TiO_2 concentrations. Yaghoubi et al. demonstrated that a polycarbonate substrate with a self-cleaning TiO_2 layer can increase mechanical endurance by improving hardness and scratch resistance (Yaghoubi et al., 2010). These results show that the self-cleaning coatings are important in making these coatings appealing in the automotive and construction industries. Chaudari et al. investigated the influence of nano TiO_2 pre-treatment on cotton fabric functional characteristics (Chaudhari et al., 2012). The pad-dry-cure process was used to treat the cotton cloth with nano TiO_2 colloid. Similarly, Ortelli et al. used the pad-dry-cure process to apply TiO_2 nanosol directly to fabrics (Ortelli et al., 2015). The cloth was soaked in TiO_2 nanosol for 3 minutes before being run through a two-roller laboratory padder, oven dried, and treated for 10 minutes at 130°C.There have also been some attempts to functionalize fabrics with ZnO in order to provide self-cleaning capabilities. The photocatalytic solution discolouration and self-cleaning of functionalized polyester fabric with ZnO nanorods were investigated by Ashraf et al. ZnO nanorods were hydrothermally deposited on polyester fabric thereby degrading stains and decolorizing dye solutions of various sorts (Ashraf et al., 2015). Due to the photocatalytic action of ZnO, they discovered that the functionalized cloth destroyed the stains and decolored the solution. The fabric withstood repeated solution discoloration, however, its stain degradability decreased with each degradation.

7.10 SENSORS ON TEXTILE

Numerous types of sensors, including heat sensors, touch sensors, pressure sensors, optical sensors, chemical sensors, olfactory sensors, etc., can be incorporated on the textile for a range of purposes (Ullah et al., 2018). For usage as lightweight, flexible, and high strain sensors that may be employed in the domains of smart clothing, health monitoring, and human motion detection, carbon-based nanomaterials such as carbon nanofibers, graphene, and CNT have been thoroughly investigated. Different methods have been used to create carbon-based nanoparticles, which are then uniformly disseminated into polymers for use as strain sensors. Both direct film casting and electrospinning processes were used to create strain sensors (Liu et al., 2016). For usage in effective performance strain sensors, carbon-based nanofibers and the woven materials they are made of have been studied. Human hairs coated with graphene have also been used to create strain sensors (Yuan et al., 2015). After carbonization, stabilization, and spray coating, silk and cotton fabrics were also employed as strain sensors (Wang et al., 2017). At the moment, plasmon-based sensors are widely used in smart textiles. It has been discovered that plasmonic sensors offer exceptional sensitivity for biological detection. Using the drawing techniques, many plasmonic OFSs can be created. A plasmon resonance concept governs how plasmonic fiber sensors function. When the phase-matching condition between the two modes occurs between them at a particular frequency, a surface plasmon mode located on a metal/dielectric interface is activated by an optical fiber core-guided mode due to resonance. The phase-matching condition is altered by changes in the refractive index of a substance on the metal layer, which causes a displacement of the spectral dip at resonance, which is recorded as a signal. For the creation of a plasmonic sensor, numerous changes are used in addition to the traditional single- or multimode optic fibers, including etching, cladding, polishing, and additional deposition of several tens of metal nanolayers. The development of plasmonic fiber sensors is challenged by a number of these techniques; however, the stack-and-draw methodology can guarantee the creation of high-quality plasmonic fiber sensors. Additionally, touch sensor fabrics that use flexible capacitors in the fiber have been created. Using a Dobby loom, a 1D sensor array made of capacitor fibers was created, which was subsequently inserted into a wool matrix (Gorgutsa et al., 2011). The touch sensor fabric was created using 15 capacitor fibers. When these capacitor fibers are touched with a finger, a change in the local current and voltage distribution occurs and is recorded to perceive the contact. To create a working electric circuit on the garment, these fiber capacitors can also be combined with additional conductive fibers or battery fibers. Applications for this characteristic can be found in safety clothing, programmable textiles, and the fashion industry. Additionally, pressure-sensitive fabrics have been created. Using a dye-coating technique, organic conductive polymers, such as poly(3,4-ethylenedioxythiophene) and poly(styrenesulfonate), were applied to the fibers for use as pressure sensors, together with a dielectric perfluoropolymer layer. The pristine nylon fibers were employed to fill the remaining spaces in the matrix while these treated fibers served as the wefts and warps. The capacitors were created at the points where the fibers converged. The fabric's capacitance varied from 0.22 to 0.63 pF with a sensitivity range of 0.98–9.80 Ncm^{-2}, when 4.9 Ncm^{-2} pressure was applied to it. Similar to this, the fabrics also contain temperature and humidity sensors (Takamatsu et al., 2012). The sensors are woven into fabrics using cutting-edge methods like photolithography and inkjet printing.

A smart leotard was also created by Tesconi et al. to track an athlete's lower body movement while they were rowing (Tesconi et al., 2007). The clothing was made up of a number of printed sensors made of a graphite/silicone elastomeric composite that tracks the position of the knee and hip when rowing. In detecting flexion-extension angles during knee movement, the sensing garment was compared to a commercial device. The measured resistance change was interpreted into kinematic output using computational techniques. Gibbs et al. created sensor pants that can detect knee and hip joint movement (Gibbs and Asada, 2005). In this case, the sensing devices are arrays of conductive textile fibers placed around joints. When compared to normal goniometer data, the sensing mechanism has been effective in tracking joint mobility. An upper body sensing garment to detect posture and movement has also been reported, with a specific focus on stroke recovery. Lorussi et al. developed a sensor glove for capturing finger movements for use in rehabilitation using a similar approach (Lorussi et al., 2004). The sensing glove is made of elastic fabric with coated patches of conductive elastomer (silicone/graphite); wide patches serve as sensors, while thin patches serve as signal carriers.

Inertial sensors are built with one or a combination of two or more accelerometers, gyroscopes, and magnetometers. These sensors, usually located within an inertial measurement unit (IMU), (Figure 7.4) the function is based on both Newton's inertia law and Newton's second law. OFSs (Figure 7.4) are developed based on the changes in light transmittance. It comprises three major components: a light source to generate the light beam, an optical fiber as the traveling tunnel for the beam, and a photodetector to receive the intensity-attenuated light beam. One can measure the bending angle of the optical fiber through simple calculation on the light intensity attenuation. The most important advantage of this method, compared with IMU and other high-resolution systems, is its immunity to electromagnetic noises (Homayounfar and Andrew, 2020).

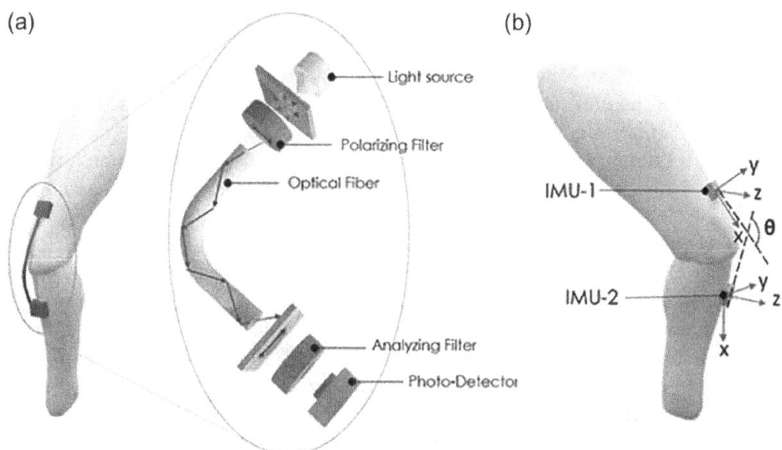

FIGURE 7.4 Schematic illustration of (a) optical fiber sensor and (b) placement of inertial sensors for human motion monitoring. (Homayounfar and Andrew, 2020.)

REFERENCES

Anbarasan, R, T Vasudevan, G Paruthimal Kalaignan, and A Gopalan. 1999. "Chemical grafting of aniline and o-toluidine onto poly (ethylene terephthalate) fiber." *Journal of Applied Polymer Science* no. 73 (1):121–128.

Ashraf, Munir, Philippe Champagne, Anne Perwuelz, Christine Campagne, and Anne Leriche. 2015. "Photocatalytic solution discoloration and self-cleaning by polyester fabric functionalized with ZnO nanorods." *Journal of Industrial Textiles* no. 44 (6):884–898.

Barthlott, Wilhelm, and Christoph Neinhuis. 1997. "Purity of the sacred lotus, or escape from contamination in biological surfaces." *Planta* no. 202 (1):1–8.

Chaudhari, Satyajeet B, Aadhar A Mandot, and Bharat H Patel. 2012. "Effect of nano TiO_2 pretreatment on functional properties of cotton fabric." *International Journal of Engineering Research and Development* no. 1 (9):24–29.

Daoud, Walid A, and John H Xin. 2004. "Low temperature sol-gel processed photocatalytic titania coating." *Journal of Sol-Gel Science and Technology* no. 29 (1):25–29.

Das, Indranee, and Goutam De. 2015. "Zirconia based superhydrophobic coatings on cotton fabrics exhibiting excellent durability for versatile use." *Scientific Reports* no. 5 (1):1–11.

Dhineshbabu, Nattanmai Raman, and Suryasarathi Bose. 2019. "UV resistant and fire retardant properties in fabrics coated with polymer based nanocomposites derived from sustainable and natural resources for protective clothing application." *Composites Part B: Engineering* no. 172:555–563.

El-Khatib, Eman M. 2012. "Antimicrobial and self-cleaning textiles using nanotechnology." *Research Journal of Textile and Apparel* no. 16:156–174.

Ferri, Ada, Maria Rosaria Plutino, and Giuseppe Rosace. 2019. Recent trends in smart textiles: Wearable sensors and drug release systems. Paper read at AIP Conference Proceedings.

Fouda, Amr, EL Saad, Salem S Salem, and Tharwat I Shaheen. 2018. "In-Vitro cytotoxicity, antibacterial, and UV protection properties of the biosynthesized Zinc oxide nanoparticles for medical textile applications." *Microbial Pathogenesis* no. 125:252–261.

Gibbs, Peter T, and H Harry Asada 2005. "Wearable conductive fiber sensors for multi-axis human joint angle measurement." *Journal of NeuroEngineering and Rehabilitation* no. 2:7.

Gorgutsa, Stephan, Jian Feng Gu, and Maksim Skorobogatiy. 2011. "A woven 2D touchpad sensor and a 1D slide sensor using soft capacitor fibers." *Smart Materials and Structures* no. 21 (1):015010.

Hiremath, Lingayya, Sura Narendra Kumar, and P Sukanya. 2018. "Development of antimicrobial smart textiles fabricated with magnetite nano particles obtained through green synthesis." *Materials Today: Proceedings* no. 5 (10):21030–21039.

Homayounfar, S Zohreh, and Trisha L Andrew. 2020. "Wearable sensors for monitoring human motion: A review on mechanisms, materials, and challenges." *SLAS Technology* no. 25 (1):9–24.

Jost, Kristy, Carlos R Perez, John K McDonough, Volker Presser, Min Heon, Genevieve Dion, and Yury Gogotsi. 2011. "Carbon coated textiles for flexible energy storage." *Energy & Environmental Science* no. 4 (12):5060–5067.

Karthik, Subramani, Palanisamy Siva, Kolathupalayam Shanmugam Balu, Rangaraj Suriyaprabha, Venkatachalam Rajendran, and Malik Maaza. 2017. "*Acalypha indica*–mediated green synthesis of ZnO nanostructures under differential thermal treatment: Effect on textile coating, hydrophobicity, UV resistance, and antibacterial activity." *Advanced Powder Technology* no. 28 (12):3184–3194.

Kirstein, Tunde. 2013. "The future of smart-textiles development: New enabling technologies, commercialization and market trends." In Tünde Kirstein (Ed.), *Multidisciplinary know-how for smart-textiles developers*, 1–25. Elsevier, Cambridge.

Lai, Haoran, Wenyue Li, Yang Zhou, Tianyu He, Ling Xu, Siyu Tian, Xiaoming Wang, Zhaoyang Fan, Zhongli Lei, and Huan Jiao. 2019. "Hydrophilically engineered polyacrylonitrile nanofiber aerogel as a soft template for large mass loading of mesoporous poly (3, 4-ethylenedioxythiophene) network on a bare metal wire for high-rate wire-shaped supercapacitors." *Journal of Power Sources* no. 441:227212.

Liu, Qiang, Ji Chen, Yingru Li, and Gaoquan Shi. 2016. "High-performance strain sensors with fish-scale-like graphene-sensing layers for full-range detection of human motions." *ACS Nano* no. 10 (8):7901–7906.

Liu, Yuyang, Xianqiong Chen, and John Haozhong Xin. 2008. "Hydrophobic duck feathers and their simulation on textile substrates for water repellent treatment." *Bioinspiration & Biomimetics* no. 3 (4):046007.

Lorussi, Federico, Alessandro Tognetti, Maurizio Tesconi, P Pastacaldi, and Danilo De Rossi. 2004. "Strain sensing fabric for hand posture and gesture monitoring." *Studies in Health Technology and Informatics* no. 108:266–270.

Mattana, Giorgio, Piero Cosseddu, Beatrice Fraboni, George G Malliaras, Juan P Hinestroza, and Annalisa Bonfiglio. 2011. "Organic electronics on natural cotton fibres." *Organic Electronics* no. 12 (12):2033–2039.

Muñoz-Bonilla, Alexandra, and Marta Fernández-García. 2015. "The roadmap of antimicrobial polymeric materials in macromolecular nanotechnology." *European Polymer Journal* no. 65:46–62.

Ortelli, Simona, Anna Luisa Costa, and Michele Dondi. 2015. "TiO_2 nanosols applied directly on textiles using different purification treatments." *Materials* no. 8 (11):7988–7996.

Pan, Shaowu, Huijuan Lin, Jue Deng, Peining Chen, Xuli Chen, Zhibin Yang, and Huisheng Peng. 2015. "Novel wearable energy devices based on aligned carbon nanotube fiber textiles." *Advanced Energy Materials* no. 5 (4):1401438.

Patil, Aravind H, Shushilkumar A Jadhav, Vikramsinh B More, Kailas D Sonawane, and Pramod S Patil. 2019. "Novel one step sonosynthesis and deposition technique to prepare silver nanoparticles coated cotton textile with antibacterial properties." *Colloid Journal* no. 81 (6):720–727.

Ramaratnam, Karthik, Volodymyr Tsyalkovsky, Viktor Klep, and Igor Luzinov. 2007. "Ultrahydrophobic textile surface via decorating fibers with monolayer of reactive nanoparticles and non-fluorinated polymer." *Chemical Communications* no. 1 (43):4510–4512.

Riaz, Shagufta, Munir Ashraf, Tanveer Hussain, Muhammad Tahir Hussain, and Ayesha Younus. 2019. "Fabrication of robust multifaceted textiles by application of functionalized TiO_2 nanoparticles." *Colloids and Surfaces A: Physicochemical and Engineering Aspects* no. 581:123799.

Richardson, Joseph J, Wenting Liao, Jincai Li, Bohan Cheng, Chenyu Wang, Taku Maruyama, Blaise L Tardy, Junling Guo, Lingyun Zhao, and Wanping Aw. 2022. "Rapid assembly of colorless antimicrobial and anti-odor coatings from polyphenols and silver." *Scientific Reports* no. 12 (1):1–8.

Salat, Marc, Petya Petkova, Javier Hoyo, Ilana Perelshtein, Aharon Gedanken, and Tzanko Tzanov. 2018. "Durable antimicrobial cotton textiles coated sonochemically with ZnO nanoparticles embedded in an in-situ enzymatically generated bioadhesive." *Carbohydrate Polymers* no. 189:198–203.

Seung, Wanchul, Manoj Kumar Gupta, Keun Young Lee, Kyung-Sik Shin, Ju-Hyuck Lee, Tae Yun Kim, Sanghyun Kim, Jianjian Lin, Jung Ho Kim, and Sang-Woo Kim. 2015. "Nanopatterned textile-based wearable triboelectric nanogenerator." *ACS Nano* no. 9 (4):3501–3509.

Shah, Mudasir Akbar, Bilal Masood Pirzada, Gareth Price, Abel L Shibiru, and Ahsanulhaq Qurashi. 2022. "Applications of nanotechnology in smart textile industry: A critical review." *Journal of Advanced Research* no. 38:55–75.

Shim, Bong Sup, Wei Chen, Chris Doty, Chuanlai Xu, and Nicholas A Kotov. 2008. "Smart electronic yarns and wearable fabrics for human biomonitoring made by carbon nanotube coating with polyelectrolytes." *Nano Letters* no. 8 (12):4151–4157.

Shishoo, Roshan. 2002. "Recent developments in materials for use in protective clothing." *International Journal of Clothing Science and Technology* no. 14 (3/4):201–215.

Singh, Mukesh Kumar, and Annika Singh. 2013. "Ultraviolet protection by fabric engineering." *Journal of Textiles* no. 2013:579129.

Synytska, Alla, Rina Khanum, Leonid Ionov, Chokri Cherif, and Cornelia Bellmann. 2011. "Water-repellent textile via decorating fibers with amphiphilic janus particles." *ACS Applied Materials & Interfaces* no. 3 (4):1216–1220.

Takamatsu, Seiichi, Takeshi Kobayashi, Norihisa Shibayama, Koji Miyake, and Toshihiro Itoh. 2012. "Fabric pressure sensor array fabricated with die-coating and weaving techniques." *Sensors and Actuators A: Physical* no. 184:57–63.

Talebian, Sepehr, Gordon G Wallace, Avi Schroeder, Francesco Stellacci, and João Conde. 2020. "Nanotechnology-based disinfectants and sensors for SARS-CoV-2." *Nature Nanotechnology* no. 15 (8):618–621.

Tao, Xiaoming. 2005. *Wearable electronics and photonics*. Elsevier, Sawston.

Tesconi, Mario, Alessandro Tognetti, E Pasquale Scilingo, Giuseppe Zupone, Nicola Carbonaro, Danilo De Rossi, Elena Castellini, and Mario Marella. 2007. Wearable sensorized system for analyzing the lower limb movement during rowing activity. Paper read at 2007 IEEE International Symposium on Industrial Electronics.

Trovato, Valentina, Eti Teblum, Yulia Kostikov, Andrea Pedrana, Valerio Re, Gilbert D Nessim, and Giuseppe Rosace. 2020. "Sol-gel approach to incorporate millimeter-long carbon nanotubes into fabrics for the development of electrical-conductive textiles." *Materials Chemistry and Physics* no. 240:122218.

Ullah, Nisar, Muhammad Mansha, Ibrahim Khan, and Ahsanulhaq Qurashi. 2018. "Nanomaterial-based optical chemical sensors for the detection of heavy metals in water: Recent advances and challenges." *TrAC Trends in Analytical Chemistry* no. 100:155–166.

Vasantharaj, Seerangaraj, Selvam Sathiyavimal, Mythili Saravanan, Palanisamy Senthilkumar, Kavitha Gnanasekaran, Muthiah Shanmugavel, Elayaperumal Manikandan, and Arivalagan Pugazhendhi. 2019. "Synthesis of ecofriendly copper oxide nanoparticles for fabrication over textile fabrics: Characterization of antibacterial activity and dye degradation potential." *Journal of Photochemistry and Photobiology B: Biology* no. 191:143–149.

Wang, Chunya, Kailun Xia, Muqiang Jian, Huimin Wang, Mingchao Zhang, and Yingying Zhang. 2017. "Carbonized silk georgette as an ultrasensitive wearable strain sensor for full-range human activity monitoring." *Journal of Materials Chemistry C* no. 5 (30):7604–7611.

Wang, Dong, Yao Lin, Yan Zhao, and Lixia Gu. 2004. "Polyacrylonitrile fibers modified by nano-antimony-doped tin oxide particles." *Textile Research Journal* no. 74 (12):1060–1065.

Wilkinson, LJ, Richard White, and James Kevin Chipman. 2011. "Silver and nanoparticles of silver in wound dressings: A review of efficacy and safety." *Journal of Wound Care* no. 20 (11):543–549.

Yaghoubi, Houman, Nima Taghavinia, and Eskandar Keshavarz Alamdari. 2010. "Self cleaning TiO_2 coating on polycarbonate: Surface treatment, photocatalytic and nanomechanical properties." *Surface and Coatings Technology* no. 204 (9–10):1562–1568.

Yang, Hongying, Sukang Zhu, and Ning Pan. 2004. "Studying the mechanisms of titanium dioxide as ultraviolet-blocking additive for films and fabrics by an improved scheme." *Journal of Applied Polymer Science* no. 92 (5):3201–3210.

Yetisen, Ali K, Hang Qu, Amir Manbachi, Haider Butt, Mehmet R Dokmeci, Juan P Hinestroza, Maksim Skorobogatiy, Ali Khademhosseini, and Seok Hyun Yun. 2016. "Nanotechnology in textiles." *ACS Nano* no. 10 (3):3042–3068.

Yu, Minghua, Guotuan Gu, Wei-Dong Meng, and Feng-Ling Qing. 2007. "Superhydrophobic cotton fabric coating based on a complex layer of silica nanoparticles and perfluorooctylated quaternary ammonium silane coupling agent." *Applied Surface Science* no. 253 (7):3669–3673.

Yu, Wen, Xiang Li, Jianxin He, Yuankun Chen, Linya Qi, Pingping Yuan, Kangkang Ou, Fan Liu, Yuman Zhou, and Xiaohong Qin. 2021. "Graphene oxide-silver nanocomposites embedded nanofiber core-spun yarns for durable antibacterial textiles." *Journal of Colloid and Interface Science* no. 584:164–173.

Yuan, Wenjing, Qinqin Zhou, Yingru Li, and Gaoquan Shi. 2015. "Small and light strain sensors based on graphene coated human hairs." *Nanoscale* no. 7 (39):16361–16365.

Yuranova, Tat'yana, Rosa Mosteo, Jayasundera Bandara, Danièle Laub, and Jolin Kiwi. 2006. "Self-cleaning cotton textiles surfaces modified by photoactive SiO_2/TiO_2 coating." *Journal of Molecular Catalysis A: Chemical* no. 244 (1–2):160–167.

Zahid, Muhammad, Giulia Mazzon, Athanassia Athanassiou, and Ilker S Bayer. 2019. "Environmentally benign non-wettable textile treatments: A review of recent state-of-the-art." *Advances in Colloid and Interface Science* no. 270:216–250.

Zhang, Daohong, Menghe Miao, Haitao Niu, and Zhixiang Wei. 2014. "Core-spun carbon nanotube yarn supercapacitors for wearable electronic textiles." *ACS Nano* no. 8 (5):4571–4579.

Zhou, Zuowan, Longsheng Chu, Wenming Tang, and Lixia Gu. 2003. "Studies on the antistatic mechanism of tetrapod-shaped zinc oxide whisker." *Journal of Electrostatics* no. 57 (3–4):347–354.

8 SNM in Catalysis and Energy Applications

8.1 PHOTOVOLTAICS/SOLAR CELLS

Undoubtedly, the rapid growth of the world's community and industrial revolution are creating a great demand for energy, while the consumption of renewable energy sources such as geothermal, biomass, wind, and photovoltaic sources is threatening the planet's survival and has negative consequences. Among the energy sources mentioned above, photovoltaics are the technology used to convert sunlight into electrical power using suitable semiconductor materials (Iqbal et al., 2022). Photovoltaic cells are classified into four major classes based on its modifications, and these classifications are referred to as generations (Jayawardena et al., 2013).

The first generations of photovoltaic solar cells are based on crystalline film technology, which uses semiconductors like Silicon (Si) and Gallium Arsenide (GaAs). GaAs is the oldest material that has been used to manufacture solar cells due to its higher efficiency, while Si is the most widely used material for commercial purposes and accounts for almost 90% of the photovoltaic solar cell industry (Sampaio and González, 2017). The main goal of second-generation photovoltaic solar cells is cost reduction, which was the primary problem with the first-generation photovoltaic solar cells (Figure 8.1). This can be accomplished using thin-film technologies by reducing the amount of material used while simultaneously improving its quality. Based on materials that performed well in the first-generation development, this modification was made to include amorphous Si, crystalline-Si, Cadmium Telluride (CdTe), and copper indium gallium selenide. The third and fourth generations of photovoltaic cells include nanostructured materials such as metal oxides, graphene, and carbon nanotubes and quantum dots by improving the performance (Luceño-Sánchez et al., 2019).

Nowadays, scientists have developed many smart photovoltaics by including different functionalities to the respective nanomaterial. As an example, Ma and coworkers have developed a smart photovoltaic device, by integrating both functions of solar cells and smart windows, based on dye-sensitized solar cells using photochromic spiropyran derivatives as photosensitizers (Ma et al., 2015). Furthermore, a method called Aerotaxy is used to grow semiconducting nanowires on gold nanoparticles and use self-assembly techniques to align the nanowires on a substrate; forming a solar cell or other electrical devices. The AuNPs replace the Si substrate on which conventional semiconductor based solar cells are built (Samuelsson, 2021). Moreover, perovskite solar cells, based on mixed (MOFs) organic-inorganic halide perovskites ABX_3 (where $A = CH_3NH_3$ or NH_2CHNH_2, $B = Pb$ or Sn, $X = Cl$, Br, or I), have received significant attention due to the increase in solar energy trapping efficiency (Chhikara and Varma, 2019). The hybrid nanosolar cells, which combine organic and inorganic nanoparticles, have been more successful in realizing practical

DOI: 10.1201/9781003366270-8

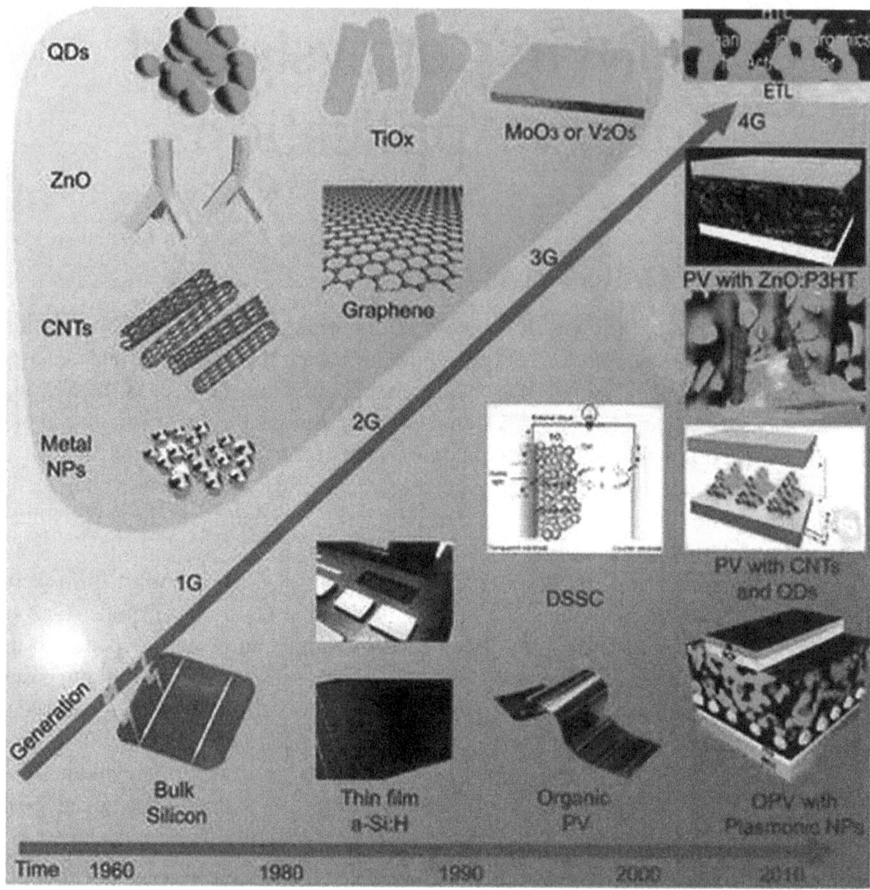

FIGURE 8.1 Four generations of photovoltaic cells along with the materials that comprise each generation. (Iqbal et al., 2022.)

solar systems. Over the last 10 years, the efficiency of perovskite-structured photovoltaic solar cells has steadily increased, reaching up to 20%.

8.2 FUEL CELLS AND BATTERIES

Redox reactions are used in fuel cells in order to convert a fuel's chemical energy into electricity (usually combining hydrogen fuel with oxygen from the air) (Winter and Brodd, 2004). They differ from most batteries in that the chemical reaction needs a constant supply of fuel and oxygen to continue. Batteries store energy in the form of metals, their oxides, or their ions; when those materials deplete, the battery stops functioning (Mekhilef et al., 2012). A fuel cell can continue to produce power indefinitely as long as it has access to oxygen and fuel. In order to produce fuel cells with the best performance, longevity, and hydrogen storage, nanotechnology is now being used to improve the materials, fuels, and production processes.

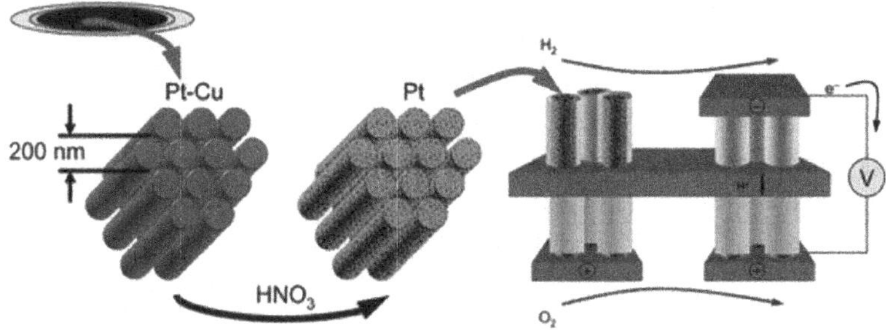

FIGURE 8.2 Template synthesis of arrays of nano fuel cells. (Lux and Rodriguez, 2006.)

Lux and co-workers have developed a method for the construction of an array of fuel cells wherein each cell is 200 nm in diameter is presented. Electrodeposition of Pt–Cu nanowires inside the cylindrical pores of an Anodisc filter membrane and the subsequent dealloying of the Cu by soaking the filter several hours in fuming nitric acid are used to develop an array of porous platinum electrodes. About 10^9 electrically isolated cylindrical porous electrodes, each 200 nm in diameter, are formed in this manner. Utilizing two arrays of porous electrodes with a polymer electrolyte membrane or an electrolyte support matrix sandwiched between, an array of nanofuel cells is produced (Lux and Rodriguez, 2006) (Figure 8.2).

Further, Pan and the team developed nanofuel and nanobiofuel cells with nanotechnology based new self-powering approaches, and the power output is sufficient to drive smart nanodevices for performing self-powered sensing. It confirms the feasibility of building smart nanosystems for applications such as environmental monitoring, biological sciences, defense technology, and even personal electronics (Pan et al., 2011).

mPhase/AlwaysReady, a company headquartered in New Jersey, is commercializing scientific findings from the field of nanotechnology (Durando, 2010). mPhase/AlwaysReady has developed a novel battery geared for sensor systems which will boost innovation initially in defense applications. The nanobattery developed by mPhase/AlwaysReady utilizes knowledge from nanotechnology and microfluidics. It makes use of a unique membrane whose nanostructured surface is extremely water-repellent. This enables a liquid electrolyte to be kept separate from the battery's electrode until energy is required. This allows the battery to be stored for an unlimited period of time before being used.

In addition, the battery is equipped with cells which can be individually activated for only those moments in which energy is needed. This is not possible in conventional batteries, where the chemical reaction cannot be interrupted until the battery has been entirely depleted. This energy on demand property gives the nanobattery a long, useful life and makes it the ideal choice for wireless light-current sensor systems. The smart nanobattery opens up new possibilities in the areas of energy storage and management. Initial concrete applications satisfy defense requirements. In the future, this technology may also be integrated in portable electronic devices. The unique nanostructured membrane can also be designed to function as a smart surface that can filter liquids. This ability opens up the potential to use the membrane design for applications such as water purification and desalination, as well as self-cleaning glass (Durando, 2010).

8.3 APPLICATION OF SNM IN RENEWABLE ENERGY SUCH AS IN CO_2 CONVERSION, WATER SPLITTING, ETC.

Because of crucial problems with energy production and usage, one of the great technological challenges of the twenty-first century is the development of renewable energy technologies. Nanotechnology and smart materials, a new promising area of research that is rapidly growing, are now considered one of the most recommended solutions to this problem. The purpose of this section is to introduce the application of nanotechnology in renewable energy systems. Shit and co-workers have designed synthesis of a novel cobalt sulfide nanoparticle grafted Porous Organic Polymer nanohybrid (CoSx@POP) and used as an active and durable water-splitting photoelectrocatalyst in the hydrogen evolution reaction (HER) (Shit et al., 2017). Recently, two-dimensional (2D) transition metal dichalcogenides (TMDs) have received great attention for solar water splitting and electrocatalysis (Andoshe et al., 2015).

Global CO_2 emissions from energy are on track for their second-largest annual increase in history. Demand for all fossil fuels was expected to skyrocket beginning in 2021. Coal demand alone is expected to rise by 60% more than all renewables combined, resulting in a 5% or 1500 Mt increase in emissions. This anticipated increase would reverse 80% of the drop in 2020, with emissions reaching 1.2% (or 400 Mt) lower than in 2019. The rate of CO_2 emissions has continued to rise, and the Earth's temperature is expected to reach a new high without intervention (Rusdan et al., 2022). In this regard, there is a great scientific attraction to SNM that are capable of CO_2 conversion. CO_2 conversion catalysts with more intricate structures than conventional heterogeneous catalysts need to be discovered (Figure 8.3).

FIGURE 8.3 Different nanomaterials in CO_2 conversion. (Wickramasinghe et al., 2021.)

Core-shell nanostructures (CSNs) are now at the core of progress in CO_2 conversion catalysis, where spherical NPs provide a controlled integration of various components and exhibit multifunctional properties (Rusdan et al., 2022). New CSNs based on cobalt (Co) catalysts have been successfully fabricated by Cui et al. to study catalytic performance of low temperature methanation. MnO-heterostructured NPs injected into porous graphitic carbon (Co/MnO@PGC) were synthesized via a single-step pyrolysis of bimetal CoMn@MOF-74. The resulting nanocomposite features an enriched Co/MnO heterointerface and exhibits excellent catalytic performance for low-temperature CO_2 methanation (Cui et al., 2021).

REFERENCES

Andoshe, Dinsefa M, Jong-Myeong Jeon, Soo Young Kim, and Ho Won Jang. 2015. "Two-dimensional transition metal dichalcogenide nanomaterials for solar water splitting." *Electronic Materials Letters* no. 11 (3):323–335.

Chhikara, Bhupender S, and Rajender S Varma. 2019. "Nanochemistry and nanocatalysis science: Research advances and future perspectives." *Journal of Materials NanoScience* no. 6 (1):1–6.

Cui, Wen-Gang, Xin-Ying Zhuang, Yan-Ting Li, Hongbo Zhang, Jing-Jing Dai, Lei Zhou, Zhenpeng Hu, and Tong-Liang Hu. 2021. "Engineering Co/MnO heterointerface inside porous graphitic carbon for boosting the low-temperature CO_2 methanation." *Applied Catalysis B: Environmental* no. 287:119959.

Durando, Ron. 2010. "New nanosurface changes battery architecture." *Advanced Coatings & Surface Technology* no. 23 (12):11–12.

Iqbal, Muhammad Aamir, Maria Malik, Wajeehah Shahid, Syed Zaheer Ud Din, Nadia Anwar, Mujtaba Ikram, and Faryal Idrees. 2022. "Materials for photovoltaics: Overview, generations, recent advancements and future prospects." In Beddiaf Zaidi and Chander Shekhar (Ed.), *Thin films photovoltaics*. IntechOpen, London.

Jayawardena, KDG Imalka, Lynn J Rozanski, Chris A Mills, Michail J Beliatis, N Aamina Nismy, and S Ravi P Silva. 2013. "'Inorganics-in-organics': Recent developments and outlook for 4G polymer solar cells." *Nanoscale* no. 5 (18):8411–8427.

Luceño-Sánchez, José Antonio, Ana María Díez-Pascual, and Rafael Peña Capilla. 2019. "Materials for photovoltaics: State of art and recent developments." *International Journal of Molecular Sciences* no. 20 (4):976.

Lux, Kenneth W, and Karien J Rodriguez. 2006. "Template synthesis of arrays of nano fuel cells." *Nano Letters* no. 6 (2):288–295.

Ma, Shengbo, Hungkit Ting, Yingzhuang Ma, Lingling Zheng, Miwei Zhang, Lixin Xiao, and Zhijian Chen. 2015. "Smart photovoltaics based on dye-sensitized solar cells using photochromic spiropyran derivatives as photosensitizers." *AIP Advances* no. 5 (5):057154.

Mekhilef, Saad, Rahman Saidur, and Azadeh Safari. 2012. "Comparative study of different fuel cell technologies." *Renewable and Sustainable Energy Reviews* no. 16 (1):981–989.

Pan, Caofeng, Jun Luo, and Jing Zhu. 2011. "From proton conductive nanowires to nanofuel cells: A powerful candidate for generating electricity for self-powered nanosystems." *Nano Research* no. 4 (11):1099–1109.

Rusdan, Nisa Afiqah, Sharifah Najiha Timmiati, Wan Nor Roslam Wan Isahak, Zahira Yaakob, Kean Long Lim, and Dalilah Khaidar. 2022. "Recent application of core-shell nanostructured catalysts for CO_2 thermocatalytic conversion processes." *Nanomaterials* no. 12 (21):3877.

Sampaio, Priscila Gonçalves Vasconcelos, and Mario Orestes Aguirre González. 2017. "Photovoltaic solar energy: Conceptual framework." *Renewable and Sustainable Energy Reviews* no. 74:590–601.

Samuelsson, Per. 2021. "Towards optical diagnostics and control in aerotaxy semiconductor nanowire growth."

Shit, Subhash Chandra, Santimoy Khilari, Indranil Mondal, Debabrata Pradhan, and John Mondal. 2017. "The design of a new cobalt sulfide nanoparticle implanted porous organic polymer nanohybrid as a smart and durable water-splitting photoelectrocatalyst." *Chemistry–A European Journal* no. 23 (59):14827–14838.

Wickramasinghe, Sameera, Jingxin Wang, Badie Morsi, and Bingyun Li. 2021. "Carbon dioxide conversion to nanomaterials: Methods, applications, and challenges." *Energy & Fuels* no. 35 (15):11820–11834.

Winter, Martin, and Ralph J Brodd. 2004. "What are batteries, fuel cells, and supercapacitors?" *Chemical Reviews* no. 104 (10):4245–4270.

9 Biogenic/Bio-Inspired Nanomaterials

9.1 USE OF BIOGENIC/BIO-INSPIRED NANOMATERIALS IN CHEMICAL ENGINEERING, TISSUE ENGINEERING, TEXTILE MANUFACTURING, NANOMEDICINE, CLINICAL DIAGNOSTICS (NANOBOTS), ELECTRONICS, ORGAN IMPLANTATIONS BIOSENSORS, BIOLOGICAL IMAGING, BIOMARKERS, AND CELL LABELLING

Bio-inspired nanomaterials, as a class of easy-to-use biomaterials, have emerged as versatile tools for a vast range of bio-medical applications such as biosensing, bio-imaging, clinical diagnostics, biocatalysis, antibacterial treatment, and biotherapy, energy applications, textile industry, etc. (Huang et al., 2015). Bioinspired nanomaterials have been synthesized in which the inspiration for synthesis is taken from nature or its components (Wu et al., 2022).

Bioinspired nanomaterials and their components have received much more attention over the last two decades, drawing inspiration from biological aspects and the field of materials technology (Gaharwar et al., 2014). After mimicking nature, these materials evolved into novel generations of materials such as bacterial-inspired, mammalian cell-inspired and virus-inspired nanosystems. Lipid, vesicles (exosomes), polysaccharides, and metallic based nanosystems are examples of commonly formed nanosystems (Kim et al., 2010). The terms "biomimetic" and "bioinspired" are used interchangeably, with a little difference in meaning. The former states directly mimic techniques or processes found in nature, whereas the latter can be direct or indirect, with a broader range of applications and greater flexibility.

Han et al. developed poly(L-lactic acid) based dual bioactive component reinforced nanofiber mats which were named as poly(L-lactic acid)/bovine serum albumin/nanohydroxyapatite (PLLA/BSA/nHAp) with dual bioactive components by combining homogeneous blending and electrospinning technology to enhance osteogenesis capability in the direction of tissue engineering (Han et al., 2021). The nanofiber mat demonstrated adequate mechanical properties and a porous structure suitable for cell growth and migration by fusing homogeneous blending and electrospinning technology, which has great potential in the use of bone repair materials. Zhou et al. reported an injectable biocompatible self-healing hydrogel adhesive with thermoresponsive reversible adhesion based on two extracellular matrix-derived biopolymers, gelatin and chondroitin sulfate, to be used as a surgical adhesive for sealing or reconnecting ruptured tissues (Zhou et al., 2021).

Due to the biocompatibility, biodegradability, and low toxicity of biogenic nanomaterials, they have received the most attention recently in the designing of drug

DOI: 10.1201/9781003366270-9

delivery systems such as protein cages (Suci et al., 2009). Drugs can be stored inside the protein cages and then delivered only into cells with the desired effects. Potential synthetic drugs can be stored in protein cages because those are hollow protein nanoparticles, such as viral capsids. Additionally, the relatively uniform amounts of drugs that can be loaded into protein cages due to their uniform cage sizes prevent the aggregation of nanoparticles (Tang et al., 2011). Uchida et al. demonstrated the versatility of using the heavy-chain ferritin (HFn) protein cage as a mineralized cage for in vitro cancer-cell-specific targeting (Uchida et al., 2006). Lin et al. evaluated ferritin nanocages as candidate nanoplatforms for multi-functional loading (Lin et al., 2011). Genetically or chemically modified ferritin nanocages can impart surface functionalities, and metal cations can be encapsulated in the interiors via metal binding sites.

Various anticancer and antimicrobial properties of bio-inspired/biogenic nanoparticles are shown in Figure 9.1.

These metallic nanoparticles (AgNPs), when treated on cancerous cells, activates the Caspase-3 pathway, which subsequently activates the different cellular process such as reactive oxygen species (ROS) generation, accumulation of autophagolysomes, endoplasmic reticulum (ER) stress generated, LDH releases, and mitochondria disrupted, and eventually leads to cell collapse, hence, cell death. The NPs degrade the peptidoglycan layer of the bacterial cell wall. These NPs disturb the electron transport chain (ETC), which eventually leads to ROS generation. They intercalate with the protein and lead to its denaturation. Furthermore, NPs also create

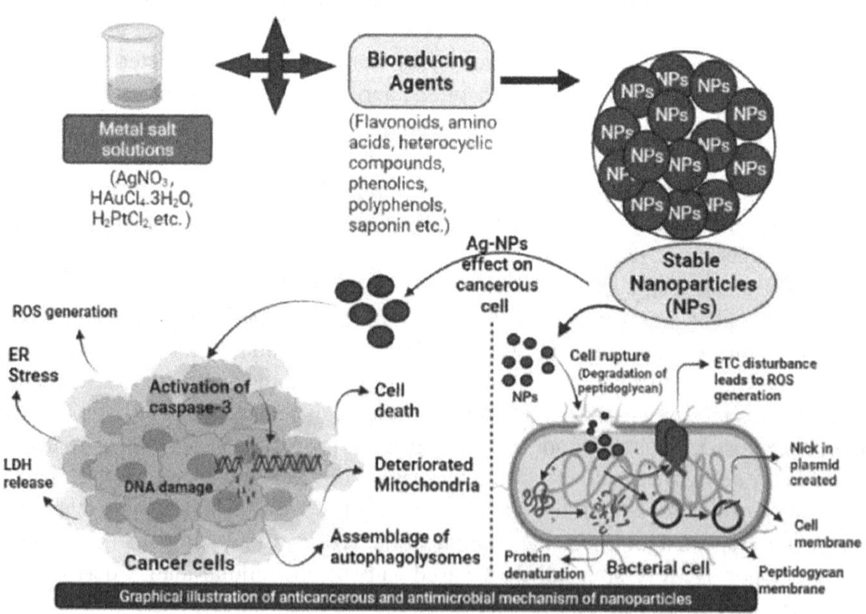

FIGURE 9.1 Metal salt solutions ($AgNO_3$, $HAuCl_4{\cdot}3H_2O$, H_2PtCl_2) after treatment with bioreducing agents (flavonoids, amino acids, heterocyclic compounds, phenolics, polyphenols, saponins, etc.) form stable nanoparticles. (Trivedi et al., 2022.)

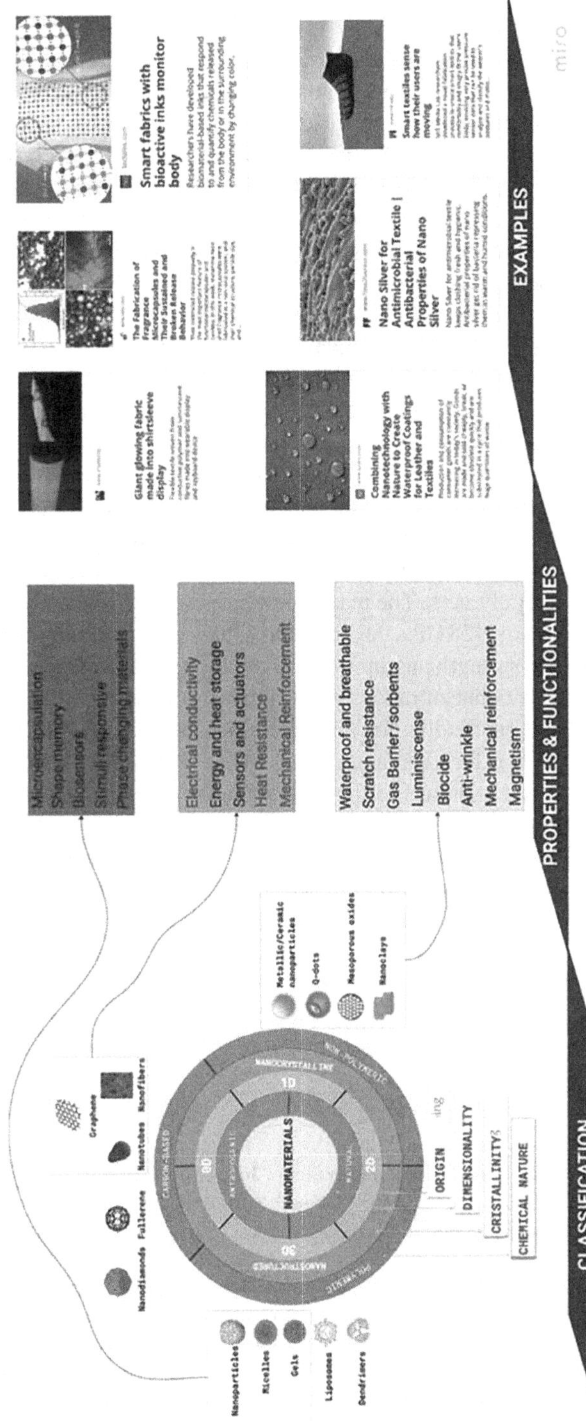

FIGURE 9.2 Scheme of the types of nanomaterials that can be used in the biotextile industry according to several properties of interest and their functionalities. (Fuentes et al., 2022.)

nicks in the plasmid. All these events inside the bacterial cell lead to its death (Haque et al., 2021).

Haque and co-workers have developed biologically synthesized silver nanoparticles and those exhibits biocompatibility and selective anticancer activity towards the cancer cell lines established through various assays. The fluorescence properties of synthesized AgNPs are observed in the near-infra-red (NIR) region (excitation: 710 nm, emission: 820 nm) in the normal and tumor-bearing C57BL6/J mice. These obtained results highlight that biologically synthesized silver nanoparticles could be an efficient cancer therapeutic and NIR based non-invasive imaging agent in the upcoming times (Haque et al., 2021).

Nano-biotechnological techniques could be used to enhance the mechanical and/or aesthetic properties of biotextiles. Mechanical reinforcement could be used to achieve the mechanical or thermal properties for particular applications, such as bags, upholstery, shoes, soles, etc (Robles et al., 2015). By adding a small amount of nanofillers and creating composites, some of which are based on ceramics (mostly silicate-based), layered nanomaterials like nanoclay, or carbon nanofillers, natural biopolymers can be strengthened (e.g., carbon fibers, carbon nanotubes, graphene, and its derivatives) (Sharma et al., 2018) (Figure 9.2). Recent research has shown that cellulose nanofibers, which are nanofibrillated by bacteria, have the potential to act as reinforcement in a variety of bio-based polymers. The maximum macroscopic Young's natural plant cellulose fibers modulus is 128 GPa, far exceeding any synthetic polymer, demonstrating the high specific strength and modulus of cellulose (Miao and Hamad, 2013). Additionally, the high dye concentration and difficult dyeing conditions make traditional dyeing techniques ineffective for polysaccharide-based fibers (Fuentes et al., 2022). Nanopigments may offer a less aggressive dyeing technique in this case. These qualities could be included in biotextiles because biofabrication in the textile industry has primarily focused on antibacterial and/or antifungal, ultraviolet light protection, and durability. AgNPs derived from black rice extract have been functionalized into cotton textiles that have demonstrated antibacterial activity against *Staphylococcus aureus* and *Escherichia coli* as well as UV protection and improved hydrophobic performance (Yu et al., 2021). Similar to this, functionalized cotton textiles made with copper oxide nanoparticles made from *Carica papaya* leaf extract showed promising antimicrobial activity against *E. coli* even after 30 wash cycles, in addition to increased hydrophobicity and tensile strength (Turakhia et al., 2020).

REFERENCES

Fuentes, Keyla M, Melissa Gómez, Hernán Rebolledo, José Miguel Figueroa, Pablo Zamora, and Leopoldo Naranjo-Briceño. 2022. "Nanomaterials in the future biotextile industry: A new cosmovision to obtain smart biotextiles." *Frontiers in Nanotechnology* no. 4:1056498.

Gaharwar, Akhilesh K, Nicholas A Peppas, and Ali Khademhosseini. 2014. "Nanocomposite hydrogels for biomedical applications." *Biotechnology and Bioengineering* no. 111 (3):441–453.

Han, Yadi, Xiaofeng Shen, Sihao Chen, Xiuhui Wang, Juan Du, and Tonghe Zhu. 2021. "A nanofiber mat with dual bioactive components and a biomimetic matrix structure for improving osteogenesis effect." *Frontiers in Chemistry* no. 9:740191.

Haque, Shagufta, Caroline Celine Norbert, Rajarshi Acharyya, Sudip Mukherjee, Muralidharan Kathirvel, and Chitta Ranjan Patra. 2021. "Biosynthesized silver nanoparticles for cancer therapy and in vivo bioimaging." *Cancers* no. 13 (23):6114.

Huang, Jiale, Liqin Lin, Daohua Sun, Huimei Chen, Dapeng Yang, and Qingbiao Li. 2015. "Bio-inspired synthesis of metal nanomaterials and applications." *Chemical Society Reviews* no. 44 (17):6330–6374.

Kim, Betty YS, James T Rutka, and Warren CW Chan. 2010. "Nanomedicine." *New England Journal of Medicine* no. 363 (25):2434–2443.

Lin, Xin, Jin Xie, Gang Niu, Fan Zhang, Haokao Gao, Min Yang, Qimeng Quan, Maria A Aronova, Guofeng Zhang, and Seulki Lee. 2011. "Chimeric ferritin nanocages for multiple function loading and multimodal imaging." *Nano Letters* no. 11 (2):814–819.

Miao, Chuanwei, and Wadood Y Hamad. 2013. "Cellulose reinforced polymer composites and nanocomposites: A critical review." *Cellulose* no. 20 (5):2221–2262.

Robles, Eduardo, Iñaki Urruzola, Jalel Labidi, and Luis Serrano. 2015. "Surface-modified nano-cellulose as reinforcement in poly (lactic acid) to conform new composites." *Industrial Crops and Products* no. 71:44–53.

Sharma, Bhasha, Parul Malik, and Purnima Jain. 2018. "Biopolymer reinforced nanocomposites: A comprehensive review." *Materials Today Communications* no. 16:353–363.

Suci, Peter A, Sebyung Kang, Mark Young, and Trevor Douglas. 2009. "A streptavidin– protein cage janus particle for polarized targeting and modular functionalization." *Journal of the American Chemical Society* no. 131 (26):9164–9165.

Tang, Zhiwen, Hong Wu, Youyu Zhang, Zhaohui Li, and Yuehe Lin. 2011. "Enzyme-mimic activity of ferric nano-core residing in ferritin and its biosensing applications." *Analytical Chemistry* no. 83 (22):8611–8616.

Trivedi, Rashmi, Tarun Kumar Upadhyay, Mohd Hasan Mujahid, Fahad Khan, Pratibha Pandey, Amit Baran Sharangi, Khursheed Muzammil, Nazim Nasir, Atiq Hassan, Nadiyah M Alabdallah, Sadaf Anwar, Samra Siddiqui, Mohd Saeed. 2022. "Recent advancements in plant-derived nanomaterials research for biomedical applications." *Processes* no. 10 (2): 338.

Turakhia, Bhavika, Madhihalli Basavaraju Divakara, Mysore Sridhar Santosh, and Sejal Shah. 2020. "Green synthesis of copper oxide nanoparticles: A promising approach in the development of antibacterial textiles." *Journal of Coatings Technology and Research* no. 17 (2):531–540.

Uchida, Masaki, Michelle L Flenniken, Mark Allen, Deborah A Willits, Bridgid E Crowley, Susan Brumfield, Ann F Willis, Larissa Jackiw, Mark Jutila, and Mark J Young. 2006. "Targeting of cancer cells with ferrimagnetic ferritin cage nanoparticles." *Journal of the American Chemical Society* no. 128 (51):16626–16633.

Wu, Gang, Xiaodan Hui, Linhui Hu, Yunpeng Bai, Abdul Rahaman, Xing-Fen Yang, and Chunbo Chen. 2022. "Recent advancement of bioinspired nanomaterials and their applications: A review." *Frontiers in Bioengineering and Biotechnology* no. 10:952523.

Yu, Wen, Xiang Li, Jianxin He, Yuankun Chen, Linya Qi, Pingping Yuan, Kangkang Ou, Fan Liu, Yuman Zhou, and Xiaohong Qin. 2021. "Graphene oxide-silver nanocomposites embedded nanofiber core-spun yarns for durable antibacterial textiles." *Journal of Colloid and Interface Science* no. 584:164–173.

Zhou, Lei, Cong Dai, Lei Fan, Yuhe Jiang, Can Liu, Zhengnan Zhou, Pengfei Guan, Yu Tian, Jun Xing, and Xiaojun Li. 2021. "Injectable self-healing natural biopolymer-based hydrogel adhesive with thermoresponsive reversible adhesion for minimally invasive surgery." *Advanced Functional Materials* no. 31 (14):2007457.

10 Pros and Cons of SNM

10.1 MANUFACTURING ADVANTAGES

A most important industry that can get benefits from nanotechnology is the manufacturing sector that will need materials like nanotubes, nanogels, nanoparticles, nanorods, etc., to manufacture various products (Figure 10.1). These materials are often stronger, more durable, and lighter than those that are not produced with the help of nanotechnology. In comparison to standard materials, nanomaterials can be engineered to have greater strength, flexibility, durability, lubricity, and electrical conductivity (Mamalis, 2007). They can also be made to be resistant to a wide range of environmental factors, including glare, moisture, temperature, corrosion, and even microbes. Today's nanotechnology enabled products, which take advantage of these properties, range from tennis rackets and baseball bats to catalysts for crude oil refining and ultrasensitive detection and identification of biological and chemical toxins (Bhushan, 2017). Nanoscale materials, structures, devices, and systems are produced at scales that are repeatable, affordable, and efficient (Taylor, 2002). Then, innovative, next-generation products that offer better performance at a lower cost and improved sustainability are created using these nanomaterials and smart devices.

10.2 ENERGY AND ELECTRONIC ADVANTAGES

Nanotechnology can actually revolutionize a lot of electronic products, procedures, and applications (Parveen et al., 2016). The areas that benefit from the continued development of smart nanotechnology when it comes to electronic products include nanotransistors, nanodiodes, organic light-emitting diodes (OLED), plasma displays, quantum computers, and many more.

Nanotechnology can also be beneficial to the energy sector. The development of more effective energy-producing, energy-absorbing, energy-harvesting and energy storage products in smaller and more efficient mini-devices is possible with this smart technology. Such items like batteries, fuel cells, and solar cells can be built smaller but can be made to be more effective with this technology (Elcock, 2007).

10.3 MEDICAL BENEFITS

In the medical world, nanotechnology is also seen as a miracle since it helps with creating what is called smart drugs and therapeutics. These drugs help to cure ailments faster and will not cause side effects compared to conventional drugs (Angeli et al., 2008). The most of the research in medicine based on nanotechnology is now focusing on areas like tissue regeneration, bone repair, immunity and even cures for such ailments like cancer, diabetes, and other life threatening diseases.

DOI: 10.1201/9781003366270-10

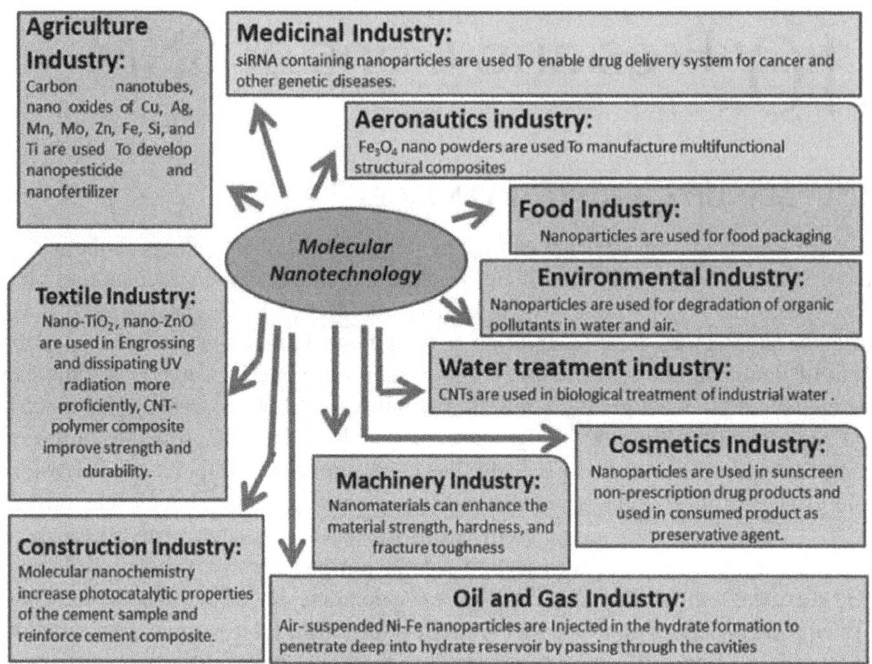

FIGURE 10.1 Applications of nanoparticle in industrial sectors. (Subhan et al., 2021.)

Most of the scientific findings exhaustively portray newly developed strategies for preparing SNM and the potential imaging or for therapeutic applications, such as smart drug delivery, tissue engineering, wound healing, theranostics, bio-markers, bio-imaging, and so on (McNeil, 2011) (Figure 10.2). A special attention is dedicated to metallic NPs for diagnosis and therapeutic applications, due to their highly tunable unique plasmonic properties (Ragni et al., 2017).

10.4 ENVIRONMENTAL EFFECTS AND ECONOMIC ISSUES

Engineered nanomaterials are rapidly becoming commonplace in cosmetics, food packaging, drug delivery systems, therapeutics, biosensors, and other applications (Ray et al., 2009). Nanomaterials are widely used in a variety of commercial products such as antimicrobial coatings, wound dressing, detergents because their size scale is similar to that of biological macromolecules and because of their antibacterial and odor-fighting properties (Ray et al., 2009). As a result, the population exposed to nanomaterials grows as their application expands. Despite the obvious benefits of nanomaterials' power, there are unanswered questions about how nanoparticles used in everyday life may affect the environment. One of the critical issues that must be addressed in the near future before mass production of nanomaterials is their toxicity to humans and environmental impact. There is considerable debate about how the novel properties of nanomaterials may cause adverse biological effects, including toxicity. What cellular responses will occur when nanoparticles undergo biodegradation

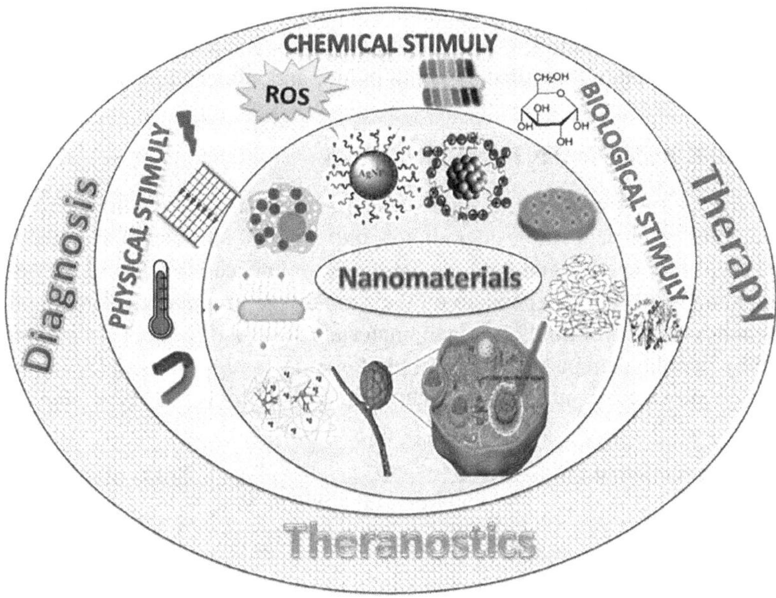

FIGURE 10.2 Classification and biomedical applications of smart nanomaterials as a function of their nanostructure. (Aflori, 2021.)

in the cellular environment? Biodegraded nanoparticles, for example, may accumulate within cells and cause intracellular changes such as organelle disruption or gene mutations. Among the critical questions are: (1) Are nanomaterials more toxic than non-nanomaterials? (2) Will nanoparticles degrade into more toxic forms in the environment? (Ray et al., 2009). Before nanomaterials can be used in everyday activities, nanotoxicology research must uncover and understand how nanomaterials influence the environment so that undesirable properties can be avoided.

This section discusses recent developments in nanomaterial toxicity and environmental impact to address concerns about potential environmental effects of emerging nanotechnologies. Both intentional and unintentional releases, such as atmospheric emissions and solid or liquid waste streams from manufacturing facilities will result in the introduction of manufactured nanomaterials into the environment. Additionally, nanomaterials found in paints, textiles, and personal care items like sunscreen and cosmetics contribute in proportion to their use to the environmental pollution. Nanomaterials released will eventually deposit on land and water surfaces. Nanomaterials that penetrate into the soil could contaminate it and move into nearby surfaces and ground waters. Wind or rainwater runoff can carry particles from solid waste, wastewater effluents, direct discharges, or unintentional spillages to the aquatic systems (Ray et al., 2009). The largest environmental releases can be caused by spills that occur during the transportation of manufactured nanomaterials from one manufacturing location to the another and by deliberate releases for environmental purposes Because engineered nanomaterials are so small, inhalation exposure to airborne particles made of nanomaterials with diameters ranging from a few nanometers to several micrometers is a possibility. The agglomeration of nanomaterials into larger particles or longer fiber chains can alter their properties and have an effect on how they behave in both indoor and outdoor settings.

10.5 RISK ASSESSMENT ON HUMANS

Due to their large surface area, high surface activity, peculiar morphology, small diameters, or after-deposition degradation into smaller particles, they can deposit in the respiratory system and have nanostructure-influenced toxicity. If they display biological activity that is dependent on their nanostructure, particles created during the breakdown or comminution of nanomaterials may also pose a risk. In the lungs of healthy people, nanoparticle deposition efficiency is high; in people with asthma or chronic obstructive pulmonary disease, it is even higher (Figure 10.3). When

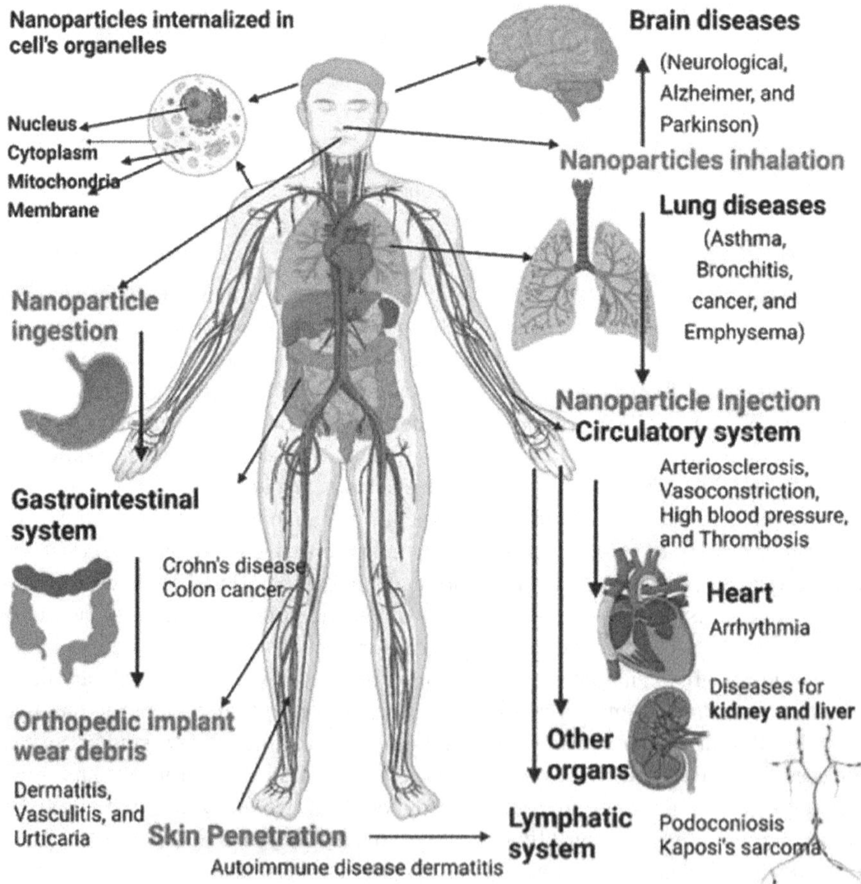

FIGURE 10.3 Schematic representation of the nanoparticle exposure routes in the human body, the organs/tissues concerned, and the diseases linked to such exposure (based on the findings of epidemiological, in vivo, and in vitro studies). (Barhoum et al., 2022.)

nanoparticles are inhaled, they settle dispersedly on the alveolar surface, which most likely causes a dispersed chemo-attractant signal and lowers the recognition and alveolar macrophage responses.

The deposition of 20 nm particles is 2.7 times greater than that of 100 nm particles and 4.3 times greater than that of 200 nm particles, according to Stahlhofen et al. (1989). According to research by Kreyling and team, patients with asthma or chronic obstructive pulmonary disease experience higher deposition efficiencies than healthy people, possibly as a result of a decline in clearance capacity (Kreyling et al., 2006). They discovered that during the first 24 hours after inhalation, there was less than 25% clearance of 50 and 100 nm particles.

Both intentional and unintentional methods can expose skin to solid nanoscale particles (Ding et al., 2005, Zhang et al., 2008). The stratum corneum, a tough layer of dead keratinized cells that makes up the outer skin, is 10 μm thick and impermeable to particles, ionic compounds, and water-soluble substances. Application of lotions, creams, bandages, detergents, and socks containing silver nanomaterials are examples of intentional dermal exposure to nanoscale materials. Nanoscale ZnO and TiO_2 materials can be exposed as sunscreen ingredients or coated on fibrous materials to give them water- or stain-repellent properties. Dermal contact with anthropomorphic substances produced during the creation or combustion of nanomaterials may result in unintentional exposure (Ray et al., 2009).

The form of the compound may also be influenced by the usage circumstances. The particles may release in a specific form during routine consumer use, but under more demanding circumstances, the form may change. Textiles, for instance, go through washing, drying, and ironing. Warm or hot water and detergent may cause a greater release of nanomaterials during washing. The nanofibers in textiles will be heated and agitated during drying. Significant heat, pressure, and abrasion are used when ironing. Risks to human health and the environment as a whole associated with the development, production, usage, and disposal of new materials must be addressed in order to develop them responsibly (Ray et al., 2009).

REFERENCES

Aflori, Magdalena. 2021. "Smart nanomaterials for biomedical applications—a review." *Nanomaterials* no. 11 (2):396.

Angeli, Elena, Renato Buzio, Giuseppe Firpo, Raffaella Magrassi, Valentina Mussi, Luca Repetto, and Ugo Valbusa. 2008. "Nanotechnology applications in medicine." *Tumori Journal* no. 94 (2):206–215.

Barhoum, Ahmed, María Luisa García-Betancourt, Jaison Jeevanandam, Eman A Hussien, Sara A Mekkawy, Menna Mostafa, Mohamed M Omran, Mohga S Abdalla, and Mikhael Bechelany. 2022. "Review on natural, incidental, bioinspired, and engineered nanomaterials: History, definitions, classifications, synthesis, properties, market, toxicities, risks, and regulations." *Nanomaterials* no. 12 (2):177.

Bhushan, Bharat. 2017. "Introduction to nanotechnology." In Bharat Bhushan (Ed.), *Springer handbook of nanotechnology*, 1–19. Springer, New York City.

Ding, Lianghao, Jackie Stilwell, Tingting Zhang, Omeed Elboudwarej, Huijian Jiang, John P Selegue, Patrick A Cooke, Joe W Gray, and Fanqing Frank Chen. 2005. "Molecular characterization of the cytotoxic mechanism of multiwall carbon nanotubes and nano-onions on human skin fibroblast." *Nano Letters* no. 5 (12):2448–2464.

Elcock, Deborah. 2007. *Potential impacts of nanotechnology on energy transmission applications and needs*. Argonne, IL: Argonne National Lab (ANL).

Kreyling, Wolfgang G, Manuela Semmler-Behnke, and Winfried Möller. 2006. "Ultrafine particle–lung interactions: Does size matter?" *Journal of Aerosol Medicine* no. 19 (1):74–83.

Mamalis, AG. 2007. "Recent advances in nanotechnology." *Journal of Materials Processing Technology* no. 181 (1–3):52–58.

McNeil, Scott E. 2011. "Unique benefits of nanotechnology to drug delivery and diagnostics." In Scott E. McNeil (Ed.), *Characterization of nanoparticles intended for drug delivery*, 3–8. Springer, New York City.

Parveen, Khadeeja, Viktoria Banse, and Lalita Ledwani. 2016. Green synthesis of nanoparticles: Their advantages and disadvantages. Paper read at AIP conference proceedings.

Ragni, Roberta, Stefania Cicco, Danilo Vona, Gabriella Leone, and Gianluca M Farinola. 2017. "Biosilica from diatoms microalgae: Smart materials from bio-medicine to photonics." *Journal of Materials Research* no. 32 (2):279–291.

Ray, Paresh Chandra, Hongtao Yu, and Peter P Fu. 2009. "Toxicity and environmental risks of nanomaterials: Challenges and future needs." *Journal of Environmental Science and Health Part C* no. 27 (1):1–35.

Stahlhofen, Willi, G Rudolf, and AC James. 1989. "Intercomparison of experimental regional aerosol deposition data." *Journal of Aerosol Medicine* no. 2 (3):285–308.

Subhan, Md Abdus, Kristi Priya Choudhury, and Newton Neogi. 2021. "Advances with molecular nanomaterials in industrial manufacturing applications." *Nanomanufacturing* no. 1 (2):75–97.

Taylor, John M. 2002. New dimensions for manufacturing: A UK strategy for nanotechnology. Office of Science and Technology London (United Kingdom).

Zhang, Leshuai W, W Yu William, Vicki L Colvin, and Nancy A Monteiro-Riviere. 2008. "Biological interactions of quantum dot nanoparticles in skin and in human epidermal keratinocytes." *Toxicology and Applied Pharmacology* no. 228 (2):200–211.

11 Can We Imagine Tomorrow Without SNM?

11.1 FUTURE OF SNM

Herein, a systematic overview of some of the newest and major advances in developing sustainable SNM is presented. This chapter exhaustively portrays newly developed strategies and future of SNM systems for wide range of applications (Sohail et al., 2019.)

The electronic transistors and microchips are the foundations of modern electronics, and they have shrunk in size since their invention in the 1940s. Today, a single chip can contain up to 5 billion transistors (Van Zant, 2014). However, in order to maintain this progress, we must be able to create circuits on an extremely small, nanometer scale. Because a nanometer (nm) is one billionth of a meter, this type of engineering entails manipulating the individual atoms. For example, we can accomplish this by directing an electron beam at a material or vaporizing it and depositing the resulting gaseous atoms layer by layer onto a base (Meindl, 2008). The real challenge is to use such techniques consistently to produce working nanoscale devices. Because the physical properties of matter, such as melting point, electrical conductivity, and chemical reactivity, change dramatically at the nanoscale, shrinking a device can have an impact on its performance. However, if we can master this technology, we will be able to improve not only electronics but also many other aspects of modern life.

Wearable fitness technology allows us to monitor our health by strapping devices to our bodies. (Figure 11.1) There are even electronic tattoo prototypes that can detect our vital signs. However, by reducing the size of this technology, we could go even further by implanting or injecting tiny sensors inside our bodies (Zohar et al., 2021). This would allow doctors to personalize their treatment by capturing much more detailed information with less hassle for the patient.

The applications range from monitoring inflammation and post-surgery recovery to more exotic applications in which electronic devices interfere with our body's signals to control organ functions. Although these technologies may appear to be a thing of the distant future, multibillion-dollar healthcare companies like GlaxoSmithKline are already working on ways to develop so-called "electroceuticals" (Donaldson and Brindley, 2016). These sensors rely on newly invented nanomaterials and manufacturing techniques to reduce their size, complexity, and energy consumption (Haroun et al., 2021) Sensors with very fine features, for example, can now be printed in large quantities on flexible rolls of plastic at a low cost (Rizwan et al., 2018). This opens up the possibility of placing sensors at various points throughout critical infrastructures to ensure that

DOI: 10.1201/9781003366270-11

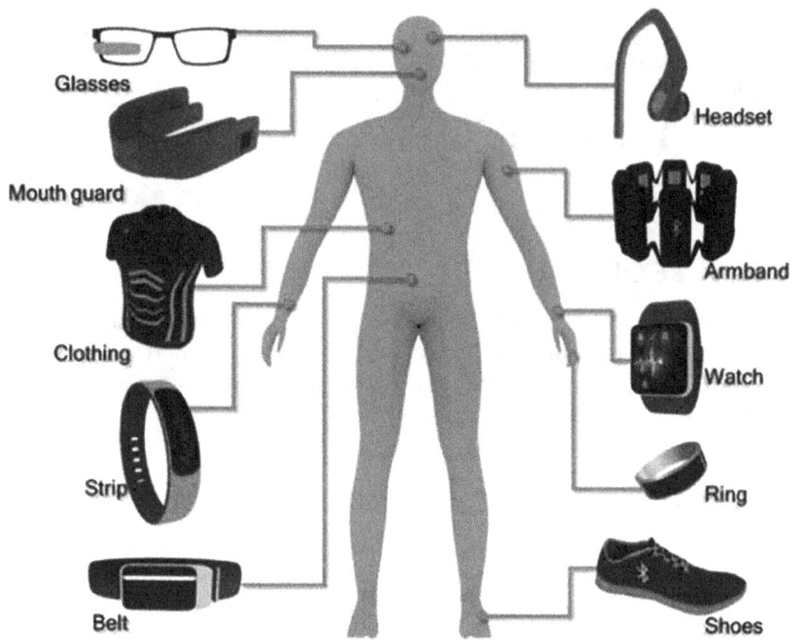

FIGURE 11.1 Portable future medical and healthcare devices worn on body parts. (Guk et al., 2019.)

everything is functioning properly. Bridges, planes, and even nuclear power plants may benefit from this (Lin, 2009).

Changing the structure of materials at the nanoscale can give them amazing properties, such as repelling water and self-cleaning. Nanocoatings or additives may even have the potential to allow materials to "heal" when damaged or worn in the future. For example, dispersing nanoparticles throughout a material allows them to migrate and fill any cracks that form. This could lead to the development of self-healing materials for everything from aircraft cockpits to microelectronics, preventing small fractures from escalating into larger, more problematic cracks (Amendola and Meneghetti, 2009).

All of these sensors will generate more data than we have ever seen before, so we will need technology to process it and identify patterns that will alert us to the problems. The same is true if we want to use "big data" from traffic sensors to help manage congestion and prevent accidents, or if we want to use statistics to more effectively allocate police resources to prevent crime (Al Najada and Mahgoub, 2016). As an example, Han and co-workers have synthesized self-sensing carbon nanotube (CNT)/cement composite for traffic monitoring. The cement composite is filled with multi-walled CNTs whose piezo resistive properties enable the detection of mechanical stresses induced by traffic flow (Han et al., 2009).

Nanotechnology is assisting in the development of ultra-dense memory, which will allow us to store this wealth of data. However, it is also inspiring ultra-efficient algorithms for processing, encrypting, and communicating data without compromising its reliability. Nature provides several examples of big-data processes that are efficiently performed in real-time by tiny structures, such as parts of the eye and ear that convert external signals into information for the brain. As an example, Sarpeshkar and team have developed bionic ears. This bionic ear allows low-power analog processing in the ear followed by digitization. Researchers at MIT have developed a bionic ear processor that does the job of the digital signal processor, is small enough to be implanted, and could run on a 2 g battery needing a wireless recharge only every 2 weeks (Sarpeshkar, 2006).

Nanotexturing is to transform a flat surface into a three-dimensional one with a much larger surface area (Anguita et al., 2016). This means that there is more room for the reactions that enable energy storage or generation to occur, resulting in more efficient operation of the devices (Hou et al., 2020). In the future, nanotechnology may allow objects to harvest energy from their surroundings. New SNM and concepts are being developed that have the potential to produce energy from movement, light, temperature variations, glucose, and other sources with high conversion efficiency. As an example, carbon nanomaterials such as nanotubes and graphene have been used in fuel cells, lithium-ion batteries, and supercapacitors and charge storage capability of supercapacitors and lithium-ion batteries (Banerjee et al., 2019).

11.2 A NEW FRONTIER; THE CONVERGENCE OF NANOTECHNOLOGY IN MACHINE LEARNING AND ARTIFICIAL INTELLIGENCE

Although the use of Artificial Intelligence (AI) in nanotechnology has not been as widely adopted as in other scientific fields, this is because the latter frequently calls for more complex systems that are not always compatible with some aspects of AI (or it makes it harder to implement) (Sacha and Varona, 2013). Nevertheless, there are some emerging fields where nanotechnology and AI are combining. An intriguing point-of-care device has recently been realized by researchers at The University of Texas Rio Grande, bridging the fields of biology, optics, nanotechnology, and AI (Mohammadinejad et al., 2016).

A removable nanotextured surface that binds specifically to breast cancer cells is housed in a microfluidic channel. It can be unbound and imaged after removal. Even though that portion is not particularly novel, machine learning makes it interesting. A machine learning algorithm is combined with segmented imaging to automatically determine whether a cell is cancerous or not by comparing it in real-time to historical data on cell size, shape, and uniformity (for both healthy and cancerous cells) (Nissim et al., 2021).

Scanning probe microscopy (SPM) is another area of imaging that has benefited from AI. This technique has historically had problems with resolution, particularly when attempting to image and manipulate samples at the nanoscale (Gordon and Moriarty, 2020). Collecting all of the various tip-to-sample interactions manually is not the easiest task because it depends on so many different parameters. The local behavior of the material being imaged can be recognized by artificial and neural networks (ANNs), which simplify the data and lowers the number of variables that need to be considered. Overall, it results in an imaging system that is much more effective.

Algorithms are already used in theoretical and computational modelling to represent a material's ideal structure, ascertain its energy and properties, and predict how it will behave in the various environments. Therefore, using AI and implementing more complex algorithms and data manipulation techniques is only a logical progression.

To accurately create either an image or a moving representation of a functioning system, many different parameters in simulations need to be correlated. Similar to some experimental imaging, AI can gather this data and learn from previous systems to produce a more accurate representation of the system under consideration, for example, by minimizing the error related to the geometry or size of a material or particle (Oliver et al., 2000). This is especially helpful for nanomaterials because it can sometimes be challenging to reproduce the various effects and phenomena observed with nanomaterials.

The future of nanocomputing is that the computation carried out by nanoscale devices can also be aided by AI. These devices can currently perform a function in a variety of ways, ranging from physical operations to computational methods. Machine learning techniques can be used to create novel data representations for a variety of applications because many of these devices depend on complex physical systems (like plasmons) to enable the execution of complex computational algorithms (Bourianoff, 2003).

Naturally, there will always be sensor networks that use a variety of AI techniques. These are not, however, strictly limited to nanosensors or nano-inspired sensors because a sensor system that applies the same AI algorithms throughout its entire network will not strictly be limited to them.

11.3 CONCLUSIONS

Smart materials facilitate advancement in a variety of fields and these materials respond to various stimuli or its environment and produce dynamic and reversible changes related to its critical properties (Yoshida and Lahann, 2008). Stimuli agents are classified as light, temperature, electric, magnetic field, stress, pressure, pH, etc. SNM are the basis of diverse emerging applications, including wearable and printed electronics, complementary metal–oxide–semiconductor (CMOS) photonics, quantum computing, artificial intelligence, optogenetics, smart coatings, and thin films. On the other hand, the almost inconceivably small devices such as sensors, cameras are becoming realistic due to the multidisciplinary field of smart nanotechnology. These changes will eventually have such a significant impact that nearly every area of science and technology will likely be impacted. As a result, smart nanotechnology holds out the possibility of bringing about the biggest technological advances in history. It is widely anticipated that smart nanotechnology will continue to develop and expand over the next couple of years in many facets of life and science, and that its accomplishments will be used in medical sciences, textile industries, agriculture and food packaging industry, energy and catalytic applications and many more industries.

REFERENCES

Al Najada, Hamzah, and Imad Mahgoub. 2016. Anticipation and alert system of congestion and accidents in VANET using Big Data analysis for Intelligent Transportation Systems. Paper read at 2016 IEEE Symposium Series on Computational Intelligence (SSCI).

Amendola, Vincenzo, and Moreno Meneghetti. 2009. "Self-healing at the nanoscale." *Nanoscale* no. 1 (1):74–88.

Anguita, José V, Muhammad Ahmad, Sajad Haq, Jeremy Allam, and S Ravi P Silva. 2016. "Ultra-broadband light trapping using nanotextured decoupled graphene multilayers." *Science Advances* no. 2 (2):e1501238.

Banerjee, Joyita, Kingshuk Dutta, and Dipak Rana. 2019. "Carbon nanomaterials in renewable energy production and storage applications." In Saravanan Rajendran, Mu. Naushad, Kumar Raju, and Rabah Boukherroub (Eds.), *Emerging nanostructured materials for energy and environmental science*, 51–104. Springer, New York City.

Bourianoff, George. 2003. "The future of nanocomputing." *Computer* no. 36 (8):44–53.

Donaldson, Nick, and Giles S Brindley. 2016. "The historical foundations of bionics." In Robert K. Shepherd (Ed.), *Neurobionics: The biomedical engineering of neural prostheses*, 1–37.John Wiley & Sons, Inc., Hoboken, NJ.

Gordon, Oliver M, and Philip J Moriarty. 2020. "Machine learning at the (sub) atomic scale: Next generation scanning probe microscopy." *Machine Learning: Science and Technology* no. 1 (2):023001.

Guk, Kyeonghye, Gaon Han, Jaewoo Lim, Keunwon Jeong, Taejoon Kang, Eun-Kyung Lim, and Juyeon Jung. 2019. "Evolution of wearable devices with real-time disease monitoring for personalized healthcare." *Nanomaterials* no. 9 (6):813.

Han, Baoguo, Xun Yu, and Eil Kwon. 2009. "A self-sensing carbon nanotube/cement composite for traffic monitoring." *Nanotechnology* no. 20 (44):445501.

Haroun, Ahmed, Xianhao Le, Shan Gao, Bowei Dong, Tianyiyi He, Zixuan Zhang, Feng Wen, Siyu Xu, and Chengkuo Lee. 2021. "Progress in micro/nano sensors and nanoenergy for future AIoT-based smart home applications." *Nano Express* no. 2 (2):022005.

Hou, Ruilin, Bao Liu, Yinglun Sun, Lingyang Liu, Jianing Meng, Mikhael D Levi, Hengxing Ji, and Xingbin Yan. 2020. "Recent advances in dual-carbon based electrochemical energy storage devices." *Nano Energy* no. 72:104728.

Lin, Yu-Feng. 2009. Smart pipe: Nanosensors for monitoring water quantity and quality in public water systems. Illinois State Water Survey.

Meindl, James D. 2008. "Nanotechnology: Retrospect and prospect." In Peter J. Hesketh (Ed.), *BioNanoFluidic MEMS*, 1–9. Springer, New York City.

Mohammadinejad, Reza, Samaneh Karimi, Siavash Iravani, and Rajender S Varma. 2016. "Plant-derived nanostructures: Types and applications." *Green Chemistry* no. 18 (1):20–52.

Nissim, Noga, Matan Dudaie, Itay Barnea, and Natan T Shaked. 2021. "Real-time stain-free classification of cancer cells and blood cells using interferometric phase microscopy and machine learning." *Cytometry Part A* no. 99 (5):511–523.

Oliver, Nuria M, Barbara Rosario, and Alex P Pentland. 2000. "A Bayesian computer vision system for modeling human interactions." *IEEE Transactions on Pattern Analysis and Machine Intelligence* no. 22 (8):831–843.

Rizwan, Ali, Ahmed Zoha, Rui Zhang, Wasim Ahmad, Kamran Arshad, Najah Abu Ali, Akram Alomainy, Muhammad Ali Imran, and Qammer H Abbasi. 2018. "A review on the role of nano-communication in future healthcare systems: A big data analytics perspective." *IEEE Access* no. 6:41903–41920.

Sacha, Gómez Moñivas, and Pablo Varona. 2013. "Artificial intelligence in nanotechnology." *Nanotechnology* no. 24 (45):452002.

Sarpeshkar, Rahul. 2006. "Brain power-borrowing from biology makes for low power computing [bionic ear]." *IEEE Spectrum* no. 43 (5):24–29.

Sohail, Muhammad Irfan, Aisha A Waris, Muhammad Ashar Ayub, Muhammad Usman, Muhammad Zia ur Rehman, Muhammad Sabir, and Tehmina Faiz. 2019. "Environmental application of nanomaterials: A promise to sustainable future." In Sandeep Kumar Verma and Ashok Kumar Das (Eds.), *Comprehensive analytical chemistry*, 1–54. Elsevier, Amsterdam.

Van Zant, Peter. 2014. *Microchip fabrication*. McGraw-Hill Education, New York City.

Yoshida, Mutsumi, and Joerg Lahann. 2008. "Smart nanomaterials." *ACS Nano* no. 2 (6):1101–1107.

Zohar, Orr, Muhammad Khatib, Rawan Omar, Rotem Vishinkin, Yoav Y Broza, and Hossam Haick. 2021. "Biointerfaced sensors for biodiagnostics." *View* no. 2 (4):20200172.

Index